ALAN
Modern Pre

ALAN AITKEN
Modern Prestressed Concrete Design

G. S. RAMASWAMY
Chief Technical Adviser,
United Nations Industrial Development Organization,
Caribbean Research Institute, Trinidad

Modern Prestressed Concrete Design

PITMAN

PITMAN PUBLISHING LIMITED
39 Parker Street, London WC2B 5PB

Associated Companies
Copp Clark Ltd, Toronto
Fearon-Pitman Publishers Inc, San Francisco
Pitman Publishing New Zealand Ltd, Wellington
Pitman Publishing Pty Ltd, Melbourne

© G. S. Ramaswamy 1976

First published in Great Britain 1976
Reprinted 1978

All rights reserved. No part of this publication may be reproduced, stored in a retrieval system, or transmitted, in any form or by any means, electronic, mechanical, photocopying, recording and/or otherwise without the prior written permission of the publishers. The paperback edition of this book may not be lent, resold, hired out or otherwise disposed of by way of trade in any form of binding or cover other than that in which it is published, without the prior consent of the publishers. This book is sold subject to the Standard Conditions of Sale of Net Books and may not be resold in the UK below the net price.

Reproduced and printed by photolithography and bound in Great Britain at The Pitman Press, Bath

ISBN 0 273 00434 4 (cased edition)
ISBN 0 273 00455 7 (paperback edition)

To my wife Seetha
and my children
Usha, Suresh and Prem

Contents

	Preface	IX
1	Introduction	1
2	Properties of materials and prestressing losses	9
3	Limit states design and classification of concrete structures	22
4	Design for flexure	27
5	Design for shear and torsion	40
6	Design of pretensioned products	54
7	Design of post-tensioned members	86
8	Transition from fully prestressed to reinforced concrete	116
9	Statically indeterminate structures	133
10	Optimum design	157
	Index	171

Preface

During the past decade, the philosophy of reinforced and prestressed concrete design has undergone revolutionary change. The traditional approach has been to regard reinforced concrete and prestressed concrete as two different materials to be dealt with in water-tight compartments. The pioneers of prestressing who recognized no halfway house between them were largely responsible for this dichotomy.

Current international thinking has, however, veered round to the view that it is far more logical to regard fully prestressed concrete and reinforced concrete as the two ends of the same spectrum, the transition from one to the other taking place through several intermediate degrees of prestressing. This line of thinking was triggered by the labours of the CEB (Comité Européen Du Béton) and FIP (Fédération Internationale de la Précontrainte) which culminated in the publication of the International Recommendations for the Design of Concrete Structures which appeared in 1970. For the first time, the transition from fully prestressed concrete to reinforced concrete was clearly delineated and structures systematically categorized (class 1 to 4), depending on the degree to which they are prestressed. Two other distinctive features of the international recommendations also need mention. Firstly, the recommendations are written in SI units. Secondly, a semi-probabilistic approach to safety has been introduced involving new concepts such as Limit States, Characteristic Strengths and Partial Safety Factors. The international recommendations are intended to serve as the general framework within which national codes are to be written. Details are left open to be spelt out to suit local needs and conditions.

The Code of Practice for the Structural Use of Concrete (CP 110: Parts 1, 2 and 3: November 1972) is the first national code to be written in the format of the international recommendations. Taking note of the CEB-FIP recommendations, it is written in SI units and presents a unified treatment of reinforced and prestressed concrete and for this reason is often referred to as the Unified Code. Being modelled on the international recommendations, the Unified Code has also incorporated the concepts of Limit States, Characteristic Strengths and Partial Safety Factors. The Cement and Concrete Association has brought out a *Handbook on the Unified Code for Structural Concrete* to provide background material and clarifications on the provisions of the Code. The author has drawn heavily on these documents and has acknowledged his indebtedness to them at appropriate places. Readers are well-advised to use the Code and the handbook available from the British Standards Institution and the Cement and Concrete Association respectively as collateral references for gaining a full understanding of the text.

The book is written with the object of providing a clear presentation of the new approach to prestressed concrete design embodied in the Unified Code and the author believes that this is the first systematic effort to do so. As is only to be expected, class 2 and class 3 structures representing intermediate degrees of prestress which have received inadequate attention in the past have been dealt with in detail. The chapter on Optimum Design which is increasingly becoming practicable because of the availability of fast computers may be mentioned as one of the special features of the book. For simplifying calculations, it has been assumed that $1 \text{ kgf} \approx 10 \text{ N}$.

The author has made every effort to make the presentation thoroughly up to date.

He acknowledges his indebtedness to his close colleagues Mr V. S. Parameswaran, Mr Zacharia George, Dr A. G. Madhava Rao, Mr Abdul Rahman, Mr Abdul Karim, Mr A. S. Prasada Rao and Mr G. Annamalai for their helpful criticism of the manuscript. He would like to specially thank Dr P. Purushothaman and Dr A. Rajaraman for their help in developing the chapter on Optimum Design.

The permission given to him by various authors, publishers and firms to quote from their publications or to use photographs supplied by them is gratefully acknowledged.

He thanks Mr T. Arokia swamy for typing for typing the manuscript and Mr P. A. Fredric for preparing the drawings.

Madras, G. S. RAMASWAMY
India.

1
Introduction

Prestressing involves the precompression of concrete to counteract, to a desired degree, the tensile stresses that are expected to occur in it in service. The village cartwright who heats a hoop of steel and fits it on the rim of a wheel has been practising prestressing without knowing it. Another commonplace example is the practice of winding ropes or metal bands over wooden staves of a barrel to resist hoop tension. Jackson of the USA is reported to have filed a patent in 1872 for a method of prestressing arches by tightening steel rods against artificial stones to construct vaulted roofs. In 1888, C. E. W. Doehring of Germany patented a method of reinforcing concrete with steel with initial tension applied to it before the slab was loaded. Other pioneers worthy of mention are the Norwegian Lund and the American Steiner. Most of these early efforts failed to produce a permanent precompression of a magnitude adequate to counteract tensile stresses occurring in service, because the pioneers of prestressing were not adequately aware of the losses in prestress that occur on account of shrinkage and creep of concrete and stress relaxation in the steel. Most of them used mild steel not perhaps realizing that the initial tension that can be imparted to it will be more than offset by the losses that take place subsequently. Eugene Freyssinet, the French engineer, who systematically investigated the time-dependent losses caused by shrinkage and creep used high tensile wires around 1928 succeeded in perfecting the technique of prestressing for practical application.

1.1 Definition of Prestressed Concrete

The definition of prestressed concrete as 'reinforced concrete in which there have been introduced internal stresses of such a magnitude and distribution that the stresses resulting from loads are counteracted *to a desired degree*' given in ACI 318:63 [1.1] has not become outdated by the CEB-FIP recommendations of 1970 [1.2]. The words 'to a desired degree' seem to have anticipated to a remarkable extent the current trend towards considering fully prestressed concrete and reinforced concrete as the two ends of the same spectrum with the transition from one to the other taking place through intermediate degrees of prestress.

1.2 Pretensioning and Post-tensioning

Concrete may be prestressed either by pretensioning or post-tensioning. In pretensioning, high tensile wires or strands are stretched to the required tension and anchored to bulkheads. The tendons are released after the concrete attains adequate strength. The bond between the steel and concrete which would have developed meanwhile resists shortening of the steel and hence the concrete gets precompressed. In post-tensioning, the concrete member is first cast with preformed holes through which the tendon is threaded and stretched against the hardened concrete and anchored to the ends of the member by means of grips. The tendons that are used for pretensioning and post-tensioning may consist of high tensile wires, rods or strands. The wires may be plain, crimped or dented.

1.3 Methods of Tensioning

In both pretensioning and post-tensioning, the tendons need to be stretched to the required tension. The tensioning may be carried out by chemical, electro-thermal or mechanical means.

1.4 Chemical Prestressing

Chemical prestressing is achieved by using an expanding cement which instead of shrinking expands after setting and during hardening. An expanding cement appears to have been first developed by the French engineer, Lossier [1.3]. The cement consisted of a blend of Portland cement, a blast furnace slag 'terminator' and a ground clinker having a composition that included dicalcium silicates, calcium aluminates and hard burned calcium sulphate. Since 1953, expansive concrete is reported to have been investigated in the Soviet Union for pressure pipes [1.4]. The composition of the cement consisted of mixtures of Portland cement, calcium aluminate cement and either gypsum or Plaster of Paris. Special hydrothermal methods of curing are reported to have been employed. Some work on chemical prestressing on a laboratory scale had been reported from the University of California at Berkeley [1.5]. The cement used in these studies was a blend of Portland cement of high tricalcium silicate and low tricalcium aluminate content with an expansive component made up of a ground clinker of alumino sulphate composition. The desired expansion is of the order of 0·60%. In the absence of restraint, the concrete will expand excessively and will not develop adequate strength. The presence of steel restricts the expansion and limits it to the desired extent.

Among structural members investigated for application of chemical prestressing were pressure pipes, two-way slabs and a hyperbolic paraboloid shell. Because of the many practical problems still to be overcome, such as the control of expansion, chemical prestressing has yet to find wide acceptance in practice.

1.5 Electro-thermal Prestressing

In this process, the prestressing tendons are stretched by heating them by means of an electric current. The technique is widely employed in the Soviet Union and Eastern European countries for the manufacture of flooring and roofing elements, precast trusses and poles. It is reported [1.6] that 75% of the 24 million cubic metres of precast prestressed products manufactured in the Sovient Union in 1967 were prestressed by using the electro-thermal method. The advantages that this process has over conventional techniques are that the capital investment on plant is very much less; high yield strength deformed bars with yield strengths of the order of 500 to 800 N/mm^2 may be used instead of the more expensive prestressing wire or strand; costly grips and anchoring devices are dispensed with; the use of deformed bars reduces the transmission length which is critical in short pretensioned members such as railway sleepers where the extreme stresses occur at

Fig. 1.1 Thermal prestressing plant

the rail-seat at a distance of about 50 cm from the ends. The labour required is also significantly less. Moreover, unlike hydraulic prestressing equipment, the plant does not pose any maintenance problems.

In this process, the tendons—usually high strength deformed bars—are stretched by heating them by means of a current on a heating stand of 10 to 24 m length (Fig. 1.1) which may accommodate 3 to 4 bars to be heated simultaneously. The heating stand consists of two end supports with brass contacts for clamping the two ends of the bars and a few intermediate supports. One of the clamping contacts holds the bar rigidly while the other slides freely on rollers mounted on one of the end supports to permit the elongation of the bar. The rods being heated may be arranged side by side horizontally or one above the other. The heating is done for a very short period to raise the

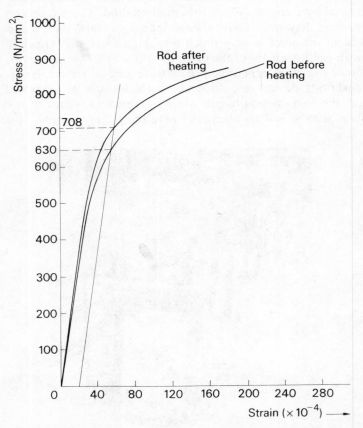

Fig. 1.2 Improvement in 0·2% proof stress of deformed bars when heated to 260°C

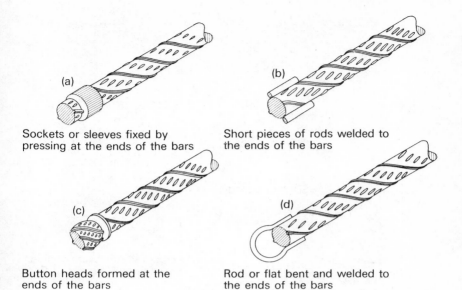

(a) Sockets or sleeves fixed by pressing at the ends of the bars

(b) Short pieces of rods welded to the ends of the bars

(c) Button heads formed at the ends of the bars

(d) Rod or flat bent and welded to the ends of the bars

Fig. 1.3 End anchorages for thermal prestressing

temperature to 250 to 400°C. Such heating for a few minutes does not anneal the steel; in fact, an improvement in the yield point is observed (Fig. 1.2). Even when the bar is hot, it is lifted off the heating stand and placed in the slots of end plates of a mould or a pretensioning bed. Inexpensive end anchorages are provided at the two ends of the bar to retain the pretension (Fig. 1.3). An electro-thermal prestressing plant patented by the Structural Engineering Research Centre, Madras [1.7] is being used in India to produce partially prestressed roofing elements such as beams and channel units (Fig. 1.4).

1.6 Mechanical Prestressing Methods

Mechanical prestressing is usually carried out by using hydraulic jacks. Detailed descriptions of systems widely employed in the UK and Europe are given by Abeles [1.8] and Leonhardt [1.9] in their books on prestressed concrete.

1.7 Reinforced versus Prestressed Concrete

It may, at this stage, be advantageous to note the similarities and differences between prestressed and reinforced concrete.

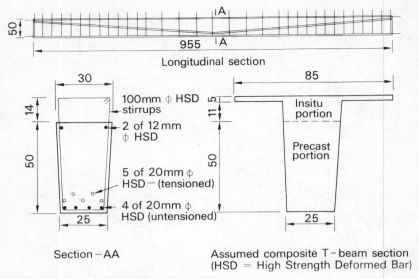

Fig. 1.4(a) Thermally prestressed beam

(1) Because of the high compression transferred by the prestressing tendons to the concrete, the compressive strength of the concrete to be used in prestressed concrete structures has to be much higher than in reinforced concrete construction.

(2) Mild steel or medium tensile steel, normally used for reinforced concrete, is unsuitable for prestressing because it cannot be stressed to an adequate extent to overcome the anticipated losses in prestress.

(3) A fully prestressed structure behaves as a homogeneous, elastic material and its behaviour before the onset of cracking is more akin to that of steel than a heterogeneous material such as reinforced concrete.

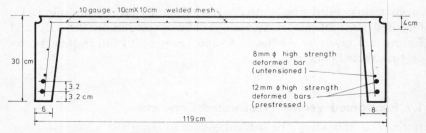

Fig. 1.4(b) Thermally prestressed channel unit

(4) A fully prestressed structure is a crack-free structure under service loads. On the other hand, a reinforced concrete structure is assumed to be cracked below the neutral axis from the very beginning. Even the cracks in a prestressed structure which open up under overloads, tend to close on the removal of load.

(5) The principal stresses in a fully prestressed beam tend to be small because of the precompression of the concrete and the reduction in vertical shear caused by the upward reactions produced by the curved tendons on the concrete. Hence, it is possible to design prestressed concrete beams with very thin webs. This leads to considerable reduction in self-weight.

(6) In both reinforced and prestressed concrete structures, the external bending moment is resisted by an internal couple, the steel being in tension and the concrete in compression. There is, however, an important difference. In a reinforced concrete beam, the lever arm remains more or less constant and as the beam is progressively loaded, the stress in the steel increases to build up the resisting moment. On the other hand, in a prestressed concrete beam, the stress in the tendon remains more or less constant and it is the change in the lever arm which contributes to the increase in the resisting moment as the load on the beam is raised.

(7) When once the prestress is overcome, the behaviour of a prestressed concrete beam does not materially differ from that of a reinforced concrete beam.

1.8 Advantages of Prestressing

Prestressed concrete structures offer the following advantages.

(1) Being made of higher strength steel and concrete, prestressed concrete is inherently superior to reinforced concrete.

(2) Prestressed concrete structures tend to be more economical than reinforced concrete structures for long spans and heavy loads.

(3) A prestressed concrete structure is a crackless structure. This is an advantage in an aggressive atmosphere and for water-retaining and other structures which call for a high degree of impermeability.

(4) Prestressed structures are lighter, partly because thin webs are practicable. The advantage is quite pronounced in long span bridges where self-weight is a dominant factor controlling the design.

(5) They deflect less because the prestressing operation causes an upward camber to start with.

(6) It is sometimes claimed that a prestressed structure is a pretested structure. What is implied is that the steel and concrete are

subjected to very high stresses during prestressing and, if the structure behaves satisfactorily at this stage, there is a reasonable assurance that it will perform equally well at other stages.

References

[1.1] *Building Code Requirements for Reinforced Concrete*, ACI 318-71, American Concrete Institute, Detroit, Michigan, 1971.

[1.2] *International Recommendations for the Design and Construction of Concrete Structures*, Principles and Recommendations, June 1970, FIP Sixth Congress, Prague, English edn, Cement and Concrete Association, London, 1971.

[1.3] Lossier, H., 'L'Autocontrainte des bétons par les cimente expansifs', *Mémoires de la Société des Ingénieurs Civils de France* (Paris) No. 3–4, March–April 1949, pp. 189–225.

[1.4] Mikhailov, V. V., 'New developments in self-stressed concrete', *Proceedings of the World Conference on Prestressed Concrete*, San Francisco, July 1957, pp. 25–1, 25–12.

[1.5] Lin, T. Y., and Klein A., 'Chemical prestressing of concrete elements using expanding cements', *Journal of the American Concrete Institute*, September 1963, Proceedings Vol. 60, No. 9, pp. 1187–1218.

[1.6] Parameswaran, V. S., Madhava Rao, A. G., and Ramachandra Murthy, D. S., 'A new technique of prestressing for economical production of precast prestressed concrete units', *Proceedings of the Symposium on Economy in Construction*, March 1974, Thiagarajar College of Engineering, Madurai, Tamil Nadu, India.

[1.7] Madhava Rao, A. G., Parameswaran, V. S., and Ramachandra Murthy, D. S., 'Precast concrete channel flooring units using high strength deformed bars prestressed by an electro-thermal method', *Indian Concrete Journal*, December 1973.

[1.8] Abeles, P. W., and Turner, F. H., *Prestressed Concrete Designer's Handbook*, Concrete Publications, London, 1962.

[1.9] Leonhardt, F., *Prestressed Concrete Design and Construction*, 2nd edn, Wilhelm Ernst und Sohn, translated into English by Amerongen, 1964.

2
Properties of materials and prestressing losses

Prestressing tendons may consist of single wires, strands or high tensile rods.

2.1 Prestressing Wires

The raw material from which wires are drawn consists of hot-rolled carbon steel rods in diameters between 5·5 and 13·5 mm with a carbon content of 0·70 to 0·85%.

The process of manufacture involves the following steps. The wire rod is given initial heat treatment known as 'patenting'. The steel is heated to a high temperature to form a solid solution of iron carbide and austenite. During gradual cooling subsequently, the ferrite separates from the austenite at about 720°C. The desired structure in the manufacture of prestressing wires is obtained by rapid cooling to 500°C followed by a delay in further cooling. For this purpose, the rods are quenched in molten salt or lead baths. The patented coil is cooled, coated and drawn through tungsten carbide dies, the dies being contained in water-cooled jackets to dissipate the heat generated. At this stage of manufacture, one obtains 'as drawn wire' conforming to BS 2691, Section 4. This mill coil wire is suitable for use in long-line prestressing beds and wire-wound pipe manufacture. However, for other uses, it is the accepted practice to specify stress-relieved wire which is obtained by straightening the 'as drawn wire' and subjecting it to a low temperature heat treatment. Stress-relieving removes the stresses induced by cold-drawing and improves the ductility and elastic properties of the wire. Stress-relieved wire which is prestraightened and supplied in coils of large diameter pays out straight and is easier to handle.

For certain applications, the losses in prestress resulting from the stress relaxation (loss of stress at constant strain) in steel need to be minimized. In such cases, stabilized wires, manufactured by the simultaneous application of a specified tensile stress and low-heat treatment, may be used. The wire is stretched to about 1% and at the same time is subjected to heat treatment at 350° to 400°C. This treatment produces a wire with more consistent stress-strain characteristics than stress-relieved wire. Moreover, the proportional limit is raised and the

10 Modern prestressed concrete design

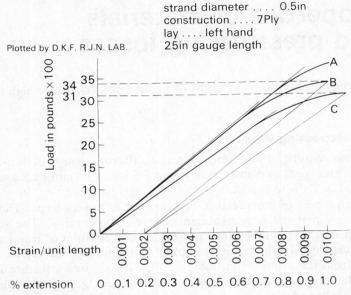

Fig. 2.1 Stress-strain curves

stress relaxation of the wire over long periods of time is reduced to one fifth of the value for stress-relieved wire. The relaxation loss of such wire at 75% of the minimum specified breaking load during 1000 hours is only 2%. Wires used as tendons are sometimes crimped or indented to improve bond. The elastic properties of stabilized, stress-relieved and 'as drawn wire' are shown compared in Fig. 2.1. (A, strand subjected to simultaneous stress heat treatment; B, strand after low temperature heat treatment i.e. stress relieved strand; C, strand, untreated.)

2.2 Prestressing Strands

Although the production of prestressing wires started in the United Kingdom as early as 1940 and the production of strands dates back only to 1956; by 1970 the production of strands exceeded that of wires. Where large prestressing forces are involved, it is advantageous to specify strands instead of single wires. Seven and nineteen wire steel strands are covered by BS 3617 and BS 4757 respectively. Strand-making, a well-known technique in rope-making, consists of a single layer or multiple layers of wire laid in helices over a central core.

Properties of materials and prestressing losses 11

Seven wire strands are the most common. They consist of a single straight wire which forms the core with six wires laid around it in a single layer, all the helices having the same pitch and direction. It is to be noted that in dealing with a strand, we are interested in the properties of the strand as a whole. Strands are also given a final low-heat treatment combined with stretching. Stabilized strands have a proportional limit at 0·10% set of 80% of the minimum guaranteed ultimate strength and a proof stress at 0·20% set of 90% of the minimum guaranteed ultimate tensile strength.

2.3 High Tensile Rods

Some prestressing systems, such as the British Macalloy System and the German Dywidag System, use high tensile rods instead of wires or strands. The diameters of the rods range from 20 to 40 mm and are covered by BS 4486. The material is generally a fully killed vacuum degassed chromium steel in billet form. The billet is heated to a controlled temperature and hot-rolled at the appropriate temperature in the cooling cycle to give a fine perlite structure. The rods are next cold-worked by stretching them to 90% of their characteristic strength. Threads on the rods are cold formed. The rods are available in length of 18 m. By the use of couplers, longer lengths of tendons become possible. The characteristic strength of rods is about 1000 N/mm^2.

2.4 Properties and Specifications

The important properties that are specified for prestressing tendons are the characteristic strength, the proof stress at 0·20% set and the percentage elongation.

Table 2.1 Specified characteristic strengths of prestressing wire

Nominal size mm	Specified characteristic strength $A_{ps}f_{pu}$ kN	Nominal cross-sectional area mm^2
2	6·34	3·14
2·65	10·3	5·5
3	12·2	7·1
3·25	14·3	8·3
4	21·7	12·6
4·5	25·7	15·9
5	30·8	19·6
7	60·4	38·5

Table 2.2 Specified characteristic strengths of prestressing strand

Number of wires	Nominal size mm	Specified characteristic strength $A_{ps}f_{pu}$ kN	Nominal cross-sectional area mm^2
7	6·4	44·5	24·5
	7·9	69·0	37·4
	9·3	93·5	52·3
	10·9	125	71·0
	12·5	165	94·2
	15·2	227	138·7
19	18	370	210
	25·4	659	423
	28·6	823	535
	31·8	979	660

Following the CEB-FIP recommendations for an International Code, the British Code CP 110 has specified the *characteristic strength* of tendons which is the ultimate strength below which not more than 5% of the test results will fall, if a number of tension specimens are tested to failure. The characteristic strength of prestressing wires, strands and high tensile alloy rods are given in Tables 2.1, 2.2 and 2.3.

The 0·20% set *proof stress* is found as indicated in Fig. 2.1. Through the point corresponding to 0·20% strain, a line is drawn parallel to the initial tangent of the stress-strain curve. The stress corresponding to

Table 2.3 Specified characteristic strengths of prestressing bars

Nominal size mm	Specified characteristic strength $A_{ps}f_{pu}$ kN	Nominal cross-sectional area mm^2
*20	325	314
22	375	380
*25	500	491
28	625	615
*32	800	804
35	950	961
*40	1250	1257

*Preferred sizes

the point where this line cuts the stress-strain curve is defined as the 0·20% set proof stress.

The minimum percentage elongation specified for wires over a gauge length of 25 cm is 4%. Similarly, a minimum percentage elongation of 4% is specified for strands over a gauge length of 24 inches (60 cm). A percentage elongation of 6% on a gauge length of $5\cdot65\sqrt{S_0}$ is specified where S_0 is the area of cross-section of the bar.

For more detailed information on the methods of manufacture and properties of prestressing wire and strands, reference may be made to manufacturers' catalogues [2.1] and a paper by Longbottom and Mallet [2.2].

2.5 Concrete

The cube strength of high-strength concrete specified for prestressed concrete work normally ranges from 30 to 60 N/mm².

Steam-curing at atmospheric pressure is usually resorted to by factories producing pretensioned products so that high early strength can be achieved and the prestress transferred within a few hours of casting, thus enabling the moulds to be released for reuse. The steam-curing cycle consists of a presteaming period of about 3 hours, a temperature rise period of about $2\frac{1}{2}$ hours, a period of about 4 hours during which the steaming is done at the constant maximum temperature (varying from 75°C to 85°C) and a cooling period of about 4 to 5 hours. The initial period, i.e., the number of hours that elapse

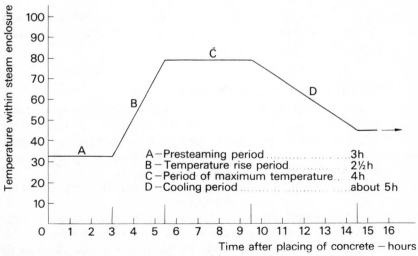

Fig. 2.2 Typical steam-curing cycle

14 Modern prestressed concrete design

after casting before steam is turned on will depend on the maximum temperature to be attained by the steam; the higher this temperature, the longer is the delay period that is specified. A typical steam-curing cycle employed by the Structural Engineering Research Centre for manufacture of railway sleepers in India is shown in Fig. 2.2.

To achieve high strength, stiff mixes are usually specified with water–cement ratios in the range of 0·40 to 0·45. Compaction of the stiff mix is done by the use of internal or form vibrators or a combination of both. With very stiff mixes, it is an advantage to use high-frequency vibrators having a frequency of 9000 cycles per minute. If very high strengths are desired, pressure may be applied in addition to vibration. For more detailed information on properties of concrete, specialized literature [2.3] on concrete technology may be consulted.

2.6 Losses in Prestress

The initial prestress imparted to the tendon at transfer becomes reduced because of the losses in tension caused by

(1) Slip in the grips or anchorages
(2) Elastic compression
(3) Friction
(4) Stress relaxation in the steel
(5) Shrinkage of the concrete
(6) Creep of the concrete
(7) Steam-curing and other miscellaneous causes

Some of the losses such as those due to slip, elastic compression and friction occur immediately on transfer, while the others take place over a period of time.

2.7 Slip in the Grips and Anchorages

Slip in the anchorages occur only in post-tensioning systems at transfer. The loss on this account is more, if the anchorage depends on wedging. This loss is practically negligible in systems employing threaded nut anchorages. The loss on account of slip is considerable, if the member is short. For example, if the member is 3 m long, the design extension will be of the order of 15 mm. If a slip of 5 mm occurs, it will account for a prestressing loss of $33\frac{1}{3}\%$. There are several methods employed in practice to overcome the loss due to slip. If the tendon is stressed and released without anchoring it at the jacking end, practically all the slip at the dead end can be eliminated. In CP 110, 10% overstressing at the time of jacking is permitted. If the anticipated slip is added to the design extension and the tendon is slightly overstressed, the design extension will remain and there is no need to allow for the slip losses.

2.8 Elastic Compression

Owing to precompression, the concrete in the neighbourhood of the stressed tendon contracts and so does the tendon, the compressive strain in the tendon being the same as that in the surrounding concrete. The loss in the prestress in the tendon on this account is found as

$$\frac{\text{stress in concrete at tendon level}}{E_c} \times E_s$$

= stress in concrete at tendon level × modular ratio.

In the above calculation, the stress in the concrete at tendon level should include the effect of the stress due to self-weight. If this stress is not included, the loss in prestress will be overestimated. For this purpose, E_s may be assumed to be 200 kN/mm^2 for wire and seven wire strands and 175 kN/mm^2 for alloy bars and 19 wire strands. The mean static modulus E_c extracted from Appendix D of CP 110 and given below in Table 2.4 may be used to compute the modular ratio.

In pretensioned members, the full loss due to elastic compression as computed above needs to be allowed for. In post-tensioned members, all the tendons are not stressed simultaneously. When there is only one tendon, the anticipated loss due to elastic compression can be added to the design extension and the tendon stretched to eliminate this loss altogether. However, if there are several tendons involved, the stretching a particular tendon causes a loss in tension in all tendons previously stretched. Suppose we have altogether N tendons; the loss in tension in the ith cable by the subsequent stretching of all cables from $i+1$th

Table 2.4 Static modulus E_c

Compressive strength in N/mm²	E_c Static modulus in kN/mm² Mean value
20	25
25	26
30	28
40	31
50	34
60	36

to N may be computed as

$$\sum_{j=i+1}^{N} f_{sj} A_{sj}\left(1+\frac{e_j y_i}{r^2}\right) \tag{2.1}$$

where f_{sj} is the stress, after all cables are stretched, in the tendon of area A_{sj} located at an eccentricity of e_j; y_i is the distance of the cable i from the centroid and A and r are the area and the radius of gyration of the section. In practice, the simple rule given in CP 110 may be followed. The rule says that the loss of prestress may be computed as the product of half the modular ratio and the stress in the concrete adjacent to the tendon averaged over the length of the tendon.

2.9 Friction

Friction between the tendon and the surface with which it is in contact causes two types of losses. Firstly, there is a loss proportional to the cumulative angle α in radians through which the cable turns. This loss is due to the curvature of the tendon. Let us consider a point P at a distance x from the jacking end O. Let the cable take several turns represented by the angles α_1, α_2 and α_3 (Fig. 2.3). $\alpha = \alpha_1 + \alpha_2 + \alpha_3$. The α values are cumulative and their signs are ignored. The curvature may be in the horizontal or vertical planes. The second type of loss is caused by the 'wobble' of the cable which is proportional to the curved length of the cable which for practical purposes is approximated by the horizontal distance x of P from O. The wobble results from the lack of rigidity of the cable or duct which causes unintentional deviations from the prescribed cable profile and gives rise to additional points of contact. The frictional losses due to curvature and wooble effect are respectively $\mu\alpha$ and kx respectively, where k is the wobble coefficient. The tension in the tendon P_x at a distance x may therefore be related to tension P_0 at the jacking end O by the relationship

$$P_x = P_0 e^{-(\mu\alpha + kx)} \tag{2.2}$$

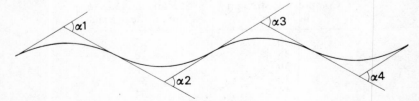

Fig. 2.3 Friction in prestressing tendons

If $\mu\alpha + kx \leq 0\cdot 20$, the relation may be approximated as

$$P_x = P_0(1 - \mu\alpha - kx) \qquad (2.3)$$

The values of μ and k prescribed in CP 110 are given in Tables 2.5 and 2.6.

Table 2.5 Values of μ

Type of contact	μ
Steel on concrete	0·55
Steel on steel	0·30
Steel on lead	0·25

Table 2.6 Values of k

Type of contact	k per m
Normal conditions	33×10^{-4}
Rigid sheath or duct formers closely supported	17×10^{-4}

2.10 Stress Relation in Steel

Stress relaxation may be defined as the loss in stress at constant strain. Normally, the manufacturers of the steel are expected to furnish the relaxation loss corresponding to the initial jacking stress in the steel at transfer based on a test conducted on the steel for 1000 hours at 20°C. In the absence of such data, the values prescribed in the Code, namely 8% and 10% of the jacking force respectively, corresponding to initial jacking stresses of 70% and 80% of the characteristic strength of the tendon, may be used. The relaxation losses in stabilized wires and strands do not exceed 2 to 3% of the initial stress. The relaxation loss varies with time according to a straight line law of the logarithmic type. CEB–FIP recommendations give the formula

$$\log\left(\frac{\text{prestress loss at time } t}{\text{initial prestress}}\right) = k_1 + k_2 \log t.$$

For stabilized steels k_1 has a more marked influence than k_2. Relaxation increases rapidly with temperature.

2.11 Shrinkage

Shrinkage is caused by the evaporation of the pore water which sets up surface tension under the action of which the concrete contracts (Fig. 2.4). The purpose of curing concrete is to delay the evaporation of water until the concrete develops a higher modulus of elasticity so that the shrinkage strain is reduced. The evaporation is primarily controlled by humidity. The more important factors controlling shrinkage are relative humidity, the composition of the concrete (as represented by its cement content), the water–cement ratio, representing the amount of water in the mix, the proportions of the member and the percentage of reinforcement. Because of the many parameters involved and the variations to which they are subject, it is almost impossible to make precise calculations for the loss in prestress due to shrinkage. Procedures are given in the CEB–FIP recommendations for taking these factors into account. But for the large majority of structures, the shrinkage strains given in Table 2.7 (extracted from CP 110) may be used. To compute the loss due to shrinkage, the shrinkage strains given are to be multiplied by E_s the modulus of elasticity of steel.

2.12 Creep

Creep is to be carefully distinguished from shrinkage. It is the time-dependent delayed strain under sustained stress. Creep may be assumed to be proportional to the stress, if the stress does not exceed one third of the cube strength. Being dependent on stress, creep strains are governed by the modulus of elasticity of concrete which, in turn, is

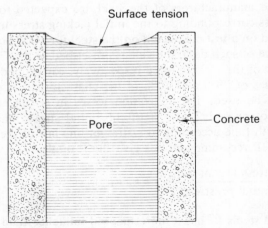

Fig. 2.4 Shrinkage due to surface tension

Table 2.7 Shrinkage of Concrete

System	Shrinkage per unit length	
	Humid exposure (90% R.H.)	Normal exposure (70% R.H.)
Pre-tensioning—transfer at 3 to 5 days after concreting	100×10^{-6}	300×10^{-6}
Post-tensioning—transfer at between 7 and 14 days after concreting	70×10^{-6}	200×10^{-6}

a function of its maturity. Young concretes creep more, and hence creep strains are dependent on the age of loading. They are also dependent on humidity conditions. The factors that influence creep may be listed as follows

(1) Relative humidity
(2) Age of concrete at loading
(3) Composition of the concrete as represented by its cement content and water–cement ratio
(4) Proportions of the member.

Elaborate methods of taking these factors into account are outlined in CEB–FIP recommendations. However, for the large majority of structures, the simple procedure given in CP 110 will suffice. For pretensioning at 3 to 5 days after concreting in humid or dry conditions of exposure, the creep coefficient may be taken as 48×10^{-6} per N/mm^2 if the cube strength of transfer is greater than 40 N/mm^2. For lower values of cube strength at transfer, the creep coefficient may be taken as $48 \times 10^{-6} \times (40/f_{ci})$ per N/mm^2. For post-tensioning at between 7 and 14 days of concreting in dry or humid conditions of exposure, where the specified cube strength of transfer exceeds 40 N/mm^2, the creep coefficient may be taken as 36×10^{-6} per N/mm^2. For lower cube strength at transfer, the creep strain may be taken as $36 \times 10^{-6} \times (40/f_{ci})$ per N/mm^2.

If the maximum stress anywhere in the section at transfer exceeds one third of the cube strength, the creep coefficient has to be increased. If the maximum stress at transfer is one half of the cube strength, the value to be used is 1·25 times the values already given. For intermediate values between one half and one third of the cube strength, linear interpolation is suggested. The loss of prestress due to creep is computed as the product of the creep strain, the stress in the concrete in the neighbourhood of the tendon and the modulus of elasticity of the steel.

The creep coefficients so far discussed correspond to the ultimate creep which will take place after a period of years. If it becomes necessary to determine creep deformation at intermediate periods, it may be assumed that half the total creep occurs in the first month after transfer and three quarters of the creep takes place in the first six months of transfer.

2.13 Interdependence of Losses

The cumulative losses arrived at by computing independently the losses due to shrinkage, creep and relaxation which are time-dependent and adding them up results in overestimating the actual loss in prestress. These losses are interdependent. For example, after a time, relaxation reduces the stress in the steel which in turn reduces the stress in the concrete. Creep loss being dependent on the stress in the concrete adjacent to the tendon, tends to get reduced. Similarly, shrinkage reduces the stress in the concrete surrounding the tendon The modulus of elasticity of concrete tends to increase with time and has to be taken into account. Where precise calculation of loss in prestress is called for, the interaction method described by Glodowski and Lorenzetti [2.4] may be used. The life of the structure is divided into about fifty steps on a logarithmic time scale and the computations are done by programming the problem for solution on a digital computer. To apply such a procedure, the relaxation losses need to be expressed as a function of time and the level of stress; the creep has also to be expressed similarly. the shrinkage needs to be expressed as a function of time. Moreover, it will also be necessary to know the time at which the various loads are applied.

2.14 Prestress Loss on account of Steam-curing

If the long-line method of pretensioning is used, the raising of the temperature of the concrete before the concrete hardens and the bond develops between the steel and the concrete results in stress relaxation in the steel whose length between anchors remains unaltered. When the concrete cools, it contracts and offsets the stress recovery in the steel. The prestress so lost needs to be taken into account. This does not happen in the individual mould method because there the temperature of the mould is raised along with that of the concrete.

2.15 Means of Reducing Losses in Post-tensioning

Friction losses in post-tensioning can be reduced by tensioning the tendon from both the ends. Restressing of tendons is also sometimes done to reduce the losses.

2.16 Importance of Accurate Assessement of Prestressing Losses

An inaccurate estimate of prestressing losses does not affect behaviour in the ultimate limit state. However, it will have a significant influence on behaviour under service loads (deflexion, camber and cracking). If the losses are overestimated, indesirable camber will result. An underestimate will lead to premature cracking.

References

[2.1] Catalogue of Richard Johnson & Nephew (Steel) Limited, U.K.

[2.2] Longbottom, K. W., and Mallet, G. P., 'Prestressing steels', *Journal of the Institution of Structural Engineers*, December 1973.

[2.3] Neville, A. M., *Properties of Concrete*, Pitman Paperbacks, London, 1970.

[2.4] Glodowski, R. J., and Lorenzetti, J. J., 'An interaction method for prestress losses in a prestressed concrete structure', *Journal of the Prestressed Concrete Institute*, March–April 1972.

3
Limit states design and classification of concrete structures

The Unified Code [3.1] which is closely modelled on the CEB–FIP recommendations for an international code [3.2] introduces the concepts of limit states design into British practice.

A structure becomes either unsafe or unserviceable, as the case may be, when one of the limit states is reached. The object of design is to ensure that there is an accepted probability that the structure or any part of it will not reach any of the limit states.

Limit states may be classified as

(1) Ultimate limit state on reaching which the structure becomes unsafe.

(2) Serviceability limit states on reaching which the structure becomes unserviceable.

For example, an *ultimate limit state* is reached when a structure or part thereof loses equilibrium when considered as a rigid body, ruptures at a section or degenerates into a mechanism. A *serviceability limit state* may be reached when a structure deflects excessively or cracks and becomes unfit for the purpose for which it is intended. Servicemay also be impaired by extensive corrosion or excessive vibration.

Limit states design is really a practical compromise between classical reliability theory and the current deterministic load factor method. It combines the merits of ultimate and working stress design. Ultimate strength design is more or less exclusively concerned with safety at overloads, and pays little or no attention to performance under working loads. On the other hand, working stress design while ensuring satisfactory behaviour under working loads gives no indication of the safety of the structure under overloads. By paying equal attention to both safety and serviceability requirements, Limit states design ensures satisfactory behaviour of the structure at all states of loading.

3.1 Characteristic loads

Characteristic loads F_k may be defined as loads which have an accepted probability of not being exceeded during the life of the structure.

Limit states design and classification of concrete structures 23

Because adequate data is not available at present on loads and their variability, CP 110 recommends that loads prescribed in CP 3, Chapter 5, be considered as characteristic loads for the time being. The *design load* is arrived at as the product of the characteristic load F_k and the partial safety factor γ_f applicable for that limit state so that

$$\text{design load} = \gamma_f \cdot F_k$$

The partial safety factor will depend on the nature and variability of the load and the importance of the limit state under consideration. Thus, a smaller partial safety factor is prescribed for the dead load and a higher factor for the imposed load because the dead load being more precisely known is subject to less variability. It is also reasonable to prescribe a higher γ_f for the ultimate limit state than for the serviceability limit state because the consequences of the structure collapsing are obviously more disastrous. The partial safety factor γ_f is introduced to account for the following

(1) Possible unusual increases in load not considered while arriving at the characteristic load

(2) Inaccurate assessment of effects of loading and stress distribution within the structure

(3) Dimensional inaccuracies of members

(4) Importance of the limit state and the probability of the load combinations prescribed

Tables 3.1 and 3.2 indicate the partial safety factors prescribed in CP 110:1972.

Nature of loading	Critical load combinations
Dead and imposed loading*	$1 \cdot 40\ G_k + 1 \cdot 60\ Q_k$
Dead and wind load†	$0 \cdot 90\ G_k + 1 \cdot 40\ W_k$
Dead, imposed and wind load	$1 \cdot 20\ G_k + 1 \cdot 20\ Q_k + 1 \cdot 20\ W_k$

G_k = Characteristic dead load
Q_k = Characteristic imposed load
W_k = Characteristic wind load
* For this loading condition, the design load $1 \cdot 0\ G_k$ with no imposed load may be more critical as in continuous beams and frames when alternate spans carry no imposed loading.
† In structures such as chimneys where the primary load is the wind load a decrease in dead load may be critical. Hence $\gamma_f < 1 \cdot 0$ is prescribed for dead load. However, in certain cases, $1 \cdot 40\ G_k + 1 \cdot 40\ W_k$ may be more critical.

Table 3.2 Load combinations for serviceability limit states (cracking and deflexion)

Nature of loading	Critical load combinations
Dead and imposed load	$1\cdot 0\, G_k + 1\cdot 0\, Q_k$
Dead and wind load	$1\cdot 0\, G_k + 1\cdot 0\, W_k$
Dead, imposed and wind load	$1\cdot 0\, G_k + 0\cdot 80\, Q_k + 0\cdot 80\, W_k$

Note: In computing deflexions of the structure or any part thereof, the imposed load needs to be so arranged that the maximum deflexion results.

The design loads prescribed above are for computing immediate deflexions. Time-dependent effects due to shrinkage and temperature will cause additional deflexions.

3.2 Characteristic Strength of Materials

The characteristic strength of materials, derived from test results on suitable specimens, is defined as that value below which not more than 5% of the results fall, i.e., 5% fractile. For normal or Gaussian distribution of test results with a standard deviation of s, the characteristic strength $f_k = f_m - 1\cdot 64\, s$, where f_m is the mean value. The characteristic strength f_k is divided by the partial safety factor γ_m for materials to arrive at the design strength which is f_k/γ_m. The Unified Code CP 110 recommends that unless otherwise stated, the characteristic strength of concrete may be taken as its 28-day strength; for steels, the characteristic strength is taken as the yield/proof strength of reinforcing steels or the ultimate load of a prestressing tendon below which not more than 5% of the test results fall. The partial safety factor γ_m is meant to account for the reduction of the strength of the material in the structure compared to the characteristic strength derived from control test specimens. Such reduction may result from the production process employed in construction. The value of γ_m prescribed for concrete and steel for the ultimate limit state are respectively $1\cdot 50$ and $1\cdot 15$. For both materials $\gamma_m = 1\cdot 0$ for the serviceability limit state of deflexion. For the serviceability limit state of cracking γ_m is taken as $1\cdot 30$ for concrete and $1\cdot 0$ for steel.

3.3 Global Safety Factor

The product $\gamma_f \cdot \gamma_m$ is known as the *global safety factor*. The reason for not lumping the two safety factors together into a single factor is that the partial safety factors permit independent allowances to be made for refinements in assessing loads and for possible future improvements in production techniques.

3.4 Classification of Concrete Structures

Following in the footsteps of the CEB–FIP recommendations, the Unified Code classifies concrete structures into four categories, depending on the degree to which they are prestressed. These four categories, Class 1 to Class 4, cover the entire spectrum of structural concrete ranging from fully prestressed to reinforced concrete. The four classes are

Class 1: Fully prestressed structures in which no tension is permitted at transfer or at working loads.

Class 2: Prestressed concrete structures in which tension is permitted but no cracking. This means that at transfer as well as under working loads, tensile stresses that develop are within the modulus of rupture which may be taken as a measure of the tensile resistance of the concrete. Although theoretically the tensile stress may be allowed to reach the modulus of rupture, the codes restrict it to about 0·80 times this value to account for the tensile stress in the concrete caused by shrinkage. Moreover, it is also usually required that the concrete does not get decompressed in the extreme tension zone at dead load $+\chi$ times the imposed load where χ is that fraction of the imposed load which is of long duration. In the CEB–FIP recommendations, loads which act for one year or more are considered to be of long duration. This requirement ensures that tension in the concrete occurs only infrequently.

Class 3: Structures in which cracking is permitted under certain working loads but crack-widths are restricted to 0·10 or 0·20 mm depending upon the type of the structure and exposure conditions. For example, for an exposed structure in an aggressive atmosphere, the crack-width is preferably restricted to 0·10 mm. For a protected structure in a normal environment, the crack-width may be 0·20 mm under working loads.

Class 4: These are reinforced concrete structures with no prestress.

Table 3.3 reproduced from [3.2] summarizes the load combinations under which different classes of structures are to be checked for decompression, incipient cracking and crack-widths. In the Table, D stands for the state of decompression which is reached when the prestress imparted to the structure is completely nullified and the stress in the bottom fibre of the concrete just becomes zero. The state of incipient cracking designated by F is reached when the tensile reaches 0·80 times the modulus of rupture.

Table 3.3 Load combinations for different classes

Class of check	Class 1	Class 2		Class 3		Class 4	
Loads to be considered	All	$0<\chi<1$	All	$0<\chi<1$	All	$0<\chi<1$	All
Very exposed members	D					0·10 (or F)	0·20*
Unprotected members	D	D	F	F (or D)	0·10	0·20	0·30
Protected members	D	D	F	F	0·20	0·30	Depends on appearance

* These figures denote the crack width in mm.

References

[3.1] *British Standard Code of Practice for the Use of Structural Concrete*, British Standards Institution, London.

[3.2] *International Recommendations for the Design and Construction of Concrete Structures, Principles and Recommendations*, June 1970, Sixth FIP Congress, Prague, English edition, Cement and Concrete Association, London.

4
Design for flexure

The following are the important stages to be considered in the design of a flexural member:

(1) Transfer
(2) Decompression
(3) Behaviour under working loads
(4) Cracking
(5) Behaviour at the ultimate limit state.

At transfer, besides the prestressing force P, the girder weight also acts and the member hogs up with an upward camber.

Gradually, the initial prestress decreases on account of various losses already mentioned in Chapter 2.

If a simply supported beam is progressively loaded, a stage is reached when the stress in the bottom fibre at any given section just becomes zero. The bending moment at that section when such a state is reached is known as the *decompression moment*. If the load is increased any further, the bottom fibre will develop tension.

At working loads, no tensile stresses are permitted in Class 1 structures. In Class 2 structures, the tensile stresses in the bottom fibre of a simply supported beam may reach the tensile strength of the concrete, suitably reduced by a factor to account for the tensile stresses caused by shrinkage. In Class 3 structures, cracking is permitted but crack-widths are limited to values prescribed in the Code. The computation of crack-widths in Class 3 beams under working loads is dealt with in Chapter 8. For computations at working loads, the effective prestressing force P_e, after deducting losses, is to be employed. The initial prestressing force P and the effective prestressing force P_e may be related by introducing the loss factor K such that $P_e = KP$.

At the ultimate limit state, a check is made in the case of Class 1 structures to see if the prestressing steel provided develops a resisting moment equal to the external bending moment. Normally, the prestressed steel provided will be adequate for this purpose in Class 1 members. In Class 2 and Class 3 structures, the prestressed steel being less, it will be found necessary to provide supplementary untensioned steel to meet ultimate limit state bending moment requirements.

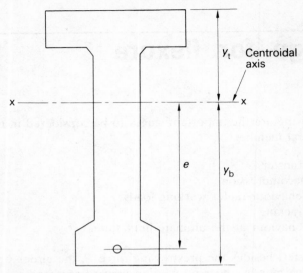

Fig. 4.1 Beam cross-section

4.1 Magnel Inequalities

The following inequalities due to Magnel [4.1] apply to prestressed Class 1 and Class 2 beams at transfer and working loads (Fig. 4.1). The four inequalities merely state that the compressive and tensile stresses in the top and bottom fibres at any cross-section are less than or equal to those prescribed in the relevant codes.

At transfer:

$$\frac{P}{A_c}\left(1+\frac{ey_b}{r^2}\right)-\frac{M_G y_b}{I_{xx}} \leq f_{cb}^t \qquad (4.1)$$

$$\frac{P}{A_c}\left(\frac{ey_t}{r^2}-1\right)-\frac{M_G y_t}{I_{xx}} \leq f_{tt}^t \qquad (4.2)$$

At working loads:

$$-\frac{KP}{A_c}\left(1+\frac{ey_b}{r^2}\right)+\frac{M_T y_b}{I_{xx}} \leq f_{tb}^w \qquad (4.3)$$

$$-\frac{KP}{A_c}\left(\frac{ey_t}{r^2}-1\right)+\frac{M_T y_t}{I_{xx}} \leq f_{ct}^w \qquad (4.4)$$

where
> y_t = distance to the top fibre from centroid
> y_b = distance to bottom fibre from centroid
> I_{xx} = Moment of inertia about x–x
> A_c = Area of the section
> M_G = Bending moment due to girder
> M_T = Total bending moment
> e = Eccentricity of prestressing force with reference to x–x.
> r = radius of gyration of the section

The fibre stresses carry two subscripts and one superscript. The superscripts t and w stand for transfer and working loads. The first subscript describes the nature of the stress (t for tension and c for compression) and the second denotes the fibre to which it relates (t for top and b for bottom). Thus f^t_{cb} stands for the compressive stress in the bottom fibre at transfer. Equations (4.1) and (4.3) relate to the bottom fibre at transfer and working loads and equations (4.2) and (4.4) apply to the top fibre at transfer and working loads. The reciprocal of the prestressing force $1/P$ and the eccentricity e may be linearly related if the equations (4.1) to (4.4) are recast as follows as suggested by Magnel.

At transfer:

$$\frac{1}{P} \geq \frac{\left(1+\dfrac{ey_b}{r^2}\right)}{A_c\left(f^t_{cb}+\dfrac{M_G y_b}{I}\right)} \tag{4.5}$$

$$\frac{1}{P} \geq \frac{\left(\dfrac{ey_t}{r^2}-1\right)}{A_c\left(f^t_{tt}+\dfrac{M_G y_t}{I}\right)} \tag{4.6}$$

At working loads:

$$\frac{1}{P} \leq \frac{K\left(1+\dfrac{ey_b}{r^2}\right)}{A_c\left(\dfrac{M_T y_b}{I}-f^w_{tb}\right)} \tag{4.7}$$

$$\frac{1}{P} \leq \frac{K\left(\dfrac{ey_t}{r^2}-1\right)}{A_c\left(\dfrac{M_T y_t}{I}-f^w_{ct}\right)} \tag{4.8}$$

30 Modern prestressed concrete design

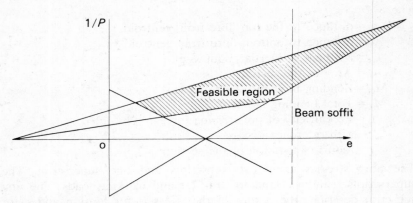

Fig 4.2 Typical Magnel chart

The advantage of establishing such linear relationships between $1/P$ and e is that the inequalities may be graphically represented (Fig. 4.2) and the feasible region found for P and e. The use of these inequalities has been illustrated in Chapter 7, Design Example 7.1.

4.2 The Pressure Line Concept

A powerful concept which is often useful in the analysis of determinate and indeterminate prestressed structures is that of the pressure line. The pressure line in a structure is found by connecting the centres of compression at all sections of the structure. We have already seen that as the load on a prestressed structure is raised, the increasing bending moments at sections are matched by resisting moments contributed by changes in the lever arm rather than the force in the tendons. This means that the centre of compression will shift in proportion to the externally applied moment. To start with, when there is no external bending moment acting, the centre of compression coincides with the point of application of the prestressing force P at all sections and hence for this condition, the pressure line coincides with the tendon profile and the resultant bending moment acting on the section is Pe due to the prestressing force. If now the section is acted upon by a positive external bending moment M (causing tension in the bottom fibres of the beam), the pressure line will shift upwards towards the centroid by an amount equal to M/P (Fig. 4.3). Hence the eccentricity of the pressure line now is $(e - M/P)$. The net moment now acting on the section is $(Pe - M)$ and this is found to be equal to the product of P and the eccentricity of the pressure line. In general, *if the position of the pressure line at a section can be located, the total stresses on the section due to prestress and the external moment can be computed as those due to an eccentric force P acting with the eccentricity of the pressure line at that section.* It needs to be clearly understood, that *the eccentricity mentioned above is that of the pressure line and not of the tendon.* If now, instead of a positive bending moment, a negative bending moment were to act on the section, the pressure line would shift downwards by an amount M/P below the tendon position so that the eccentricity of the pressure line becomes $(e + M/P)$.

This concept will now be shown applied to a few examples.

Example 4.1
Determine the eccentricity of the prestressing force in a simply supported beam, at midspan at transfer so that the stress in the top fibre is zero.

Solution
At transfer, besides the prestressing force, the bending moment M_G due to the self-weight would also act. It is clear that the pressure line should be at the bottom kern point at a distance k_b from the centroid of the section so that no tension occurs at the top. We know that $k_b = Z_t/A_c$, where Z_t is the modulus of section relating to the top fibre. Because the effect of the bending moment M_G is to shift the pressure line by a distance M_G/P above the tendon, the tendon itself must be initially located M_G/P below the bottom kern point so that after the shift the pressure line is at the bottom kern point. Hence the desired $e = [k_b + (M_G/P)]$.

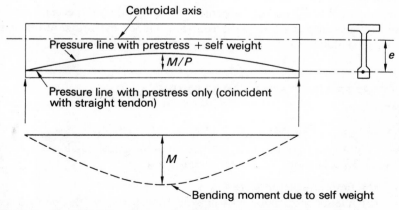

Fig. 4.3 Pressure line for a simply supported beam

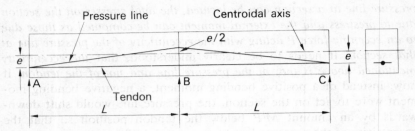

Fig. 4.4 Pressure line for a continuous beam

Example 4.2
Determine the pressure line in a continuous beam of two equal spans l (Fig. 4.4) which is prestressed by a tendon having an eccentricity e throughout.

Solution
Due to prestress the bending moment diagram is rectangular with an ordinate equal to Pe. By using the area-moment diagram theorem, the deflexion upward at B $= (Pel^2/2EI)$. Equating it to the upward deflexion at R_B,

$$\frac{R_B(2l)^3}{48EI} = \frac{Pel^2}{2EI}.$$

Or, $R_B = \dfrac{3Pe}{l} \downarrow$. The reactions $R_A = R_C = \dfrac{3}{2}\dfrac{Pe}{l} \uparrow$. The bending moment at B $= -Pe + \dfrac{3}{2}\dfrac{Pe}{l} \cdot l = \dfrac{1}{2}Pe$. Hence the eccentricity of the pressure line at B is $e/2$ above the centroid. At A and C, the eccentricity of the pressure line is e below the centroid. Hence the pressure line may now be sketched in.

ULTIMATE LIMIT STATE IN FLEXURE

4.3 Types of Failure
As a simply supported prestressed beam is progressively loaded, it will ultimately fail in one of the following ways:

(1) Fracture of the steel
(2) Yielding of the steel followed by the rise of the neutral axis and eventual failure by the crushing of the concrete

(3) Failure by crushing of the concrete not preceded by the yielding of the steel.

The manner in which the beam will fail is dependent on the amount of steel provided which is best characterized by the non-dimensional steel parameter $\lambda = \left(\dfrac{A_{ps} \cdot f_{pu}}{bdf_{cu}}\right)$, d being the effective depth. Failure by fracture of steel occurs when the steel parameter is very low. Such a brittle failure is undesirable and codes of practice exclude such a possibility by specifying that $\dfrac{A_{ps}}{bD} \times 100 \geqslant \dfrac{25}{f_{pu}}$, where D is the overall depth of the member and f_{pu} is in kg/mm². If the steel parameter is below 0·30, the beam is regarded as under-reinforced and the failure will be initiated by the yielding of the steel. The failure is ductile and there will be ample warning before the beam fails. For this reason, it should be the designer's effort to make the beam under-reinforced.

When the steel parameter increases beyond about 0·30 the beam becomes over-reinforced and failure is by crushing of the concrete (primary compression failure). The failure being brittle, over-reinforced beams are normally avoided.

4.4 Compression Stress Block and Stresses in Tensioned and Untensioned Steels

As a preliminary to the development of expressions for the ultimate flexural strength of under-reinforced beams, it is necessary to study the distribution of compressive stresses in the concrete at failure. Based on the findings of experiments, the following three assumptions will be made:

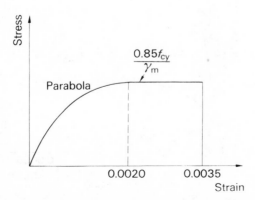

Fig. 4.5 Stress-strain curve for concrete CEB-FIP recommendations

(1) Irrespective of whether the beam is under or over-reinforced, the failure will eventually take place by the crushing of the concrete. This implies that the compression in the extreme fibre will reach the ultimate, and consequently the strain in that fibre will also be at its ultimate value of 0·0035. In the CEB–FIP recommendations, the ultimate compressive stress in the beam is taken as $0.85 f_{cy}/\gamma_m$, and the factor of 0·85 relates the cylinder strength f_{cy} to the beam or prism strength. It is to be noted that $\gamma_m = 1.5$. The stress-strain curve for concrete in compression given in the CEB–FIB recommendations consists of a rectangle and a parabolic arc (Fig. 4.5). In CP 110, the cylinder strength is replaced by cube strength, using the relationship $f_{cy} = 0.8 f_{cu}$ to give the ultimate stress in the concrete as $\dfrac{0.80 \times 0.85}{1.5} f_{cu} = 0.45 f_{cu}$. The resulting stress-strain curve for concrete in compression is given in Fig. 4.6. The shape of the stress-block will be the same as that of the stress-strain curve for concrete in compression.

(2) Strain distribution will be linear across the cross-section even at the ultimate limit state.

(3) The stress in the tensioned and untensioned steels are found from the extreme compression strain of 0·0035 given in assumption (1) and the linear strain distribution stated in assumption (2) by making use of the stress-strain curves for prestressing steel and untensioned steel (Figs 4.7 and 4.8).

In general, the procedure will consist in assuming a trial neutral axis depth and finding the total compression in the concrete and the total

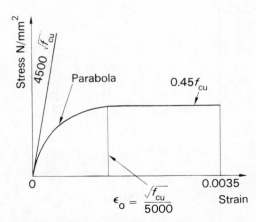

Fig. 4.6 Stress-strain curve for concrete CP 110:72

Design for flexure 35

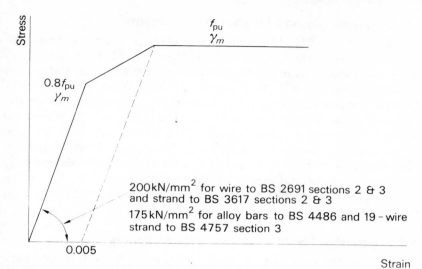

Fig. 4.7 Short term design stress-strain curve for normal and low relaxation products

tension in the steel. If these are equal, the trial neutral axis position is correct. Otherwise, the process has to be repeated, choosing other neutral axis depths until the condition stated above is satisfied. For flanged sections where the neutral axis falls outside the flange, the procedure outlined above has to be followed. Some simplifications are possible for rectangular sections and for flanged sections where the neutral axis falls within the flange.

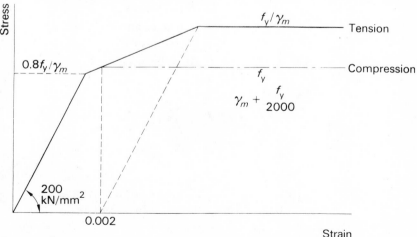

Fig. 4.8 Short-term design stress-strain curve for reinforcement

4.5 Simplified Method for Rectangular Beams

The Code CP 110:1972 permits the use of the following simplified procedure for rectangular beams. In Tables 37 and 38, the Code gives the ratio of the stress in the tendon f_{pb} at the ultimate limit state to the design strength of $0·87 f_{pu}$ in the form $f_{pb}/0·87 f_{pu}$ for different values of the steel parameter $\lambda = f_{pu} A_{ps} / f_{cu} bd$. Knowing the steel parameter, the stress f_{pb} in the steel can therefore be found. The Tables also give the ratio of (x/d), where x is the depth of the neutral axis from the top. The ultimate moment of resistance $M_u = f_{pb} \cdot A_{ps}(d - 0·5x)$, where d is the effective depth. This procedure is also applicable to flanged sections where the neutral axis lies within the flange.

Charts for rectangular beams based on the rigorous method outlined in section 4.4 are given in Part 2 of CP 110:1972.

4.6 Alternative Method for Rectangular Beams

Tests on bonded under-reinforced prestressed concrete beams have shown that the stress in the tensioned steel almost always reaches its ultimate strength. Based on a statistical analysis of available test results relating to bonded prestressed concrete beams, Ramaswamy and Narayana [4.2] have derived the following relationship between the lever arm factor j and the steel parameter λ applicable to both under-reinforced and over-reinforced beams.

$$j = e^{-0·771\lambda} \tag{4.9}$$

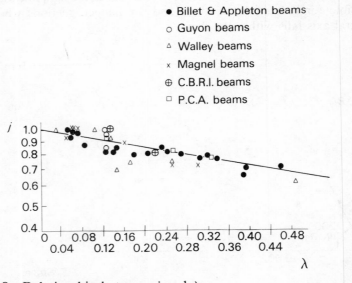

Fig. 4.9 Relationship between j and λ

Or, as may be seen from Fig. 4.9, log j and λ are linearly related when plotted. Expanding the right-hand side of equation (4.9) and retaining only one term,

$$j = \left(1 - 0.771 \frac{A_{ps}f_{pu}}{bdf_{cu}}\right) \quad (4.10)$$

The lever arm jd is the distance between the centre of compression and the centre of tension (Fig. 4.10). Noting that the stress in steel will reach its ultimate strength, the ultimate flexural strength is given by

$$M_u = A_{ps} \frac{f_{pu}d}{\gamma_m}\left(1 - 0.771 \frac{A_{ps}f_{pu}}{bdf_{cu}}\right) \quad (4.11)$$

It is interesting to note, in passing, that the formula given in ACI 318:1971 has the same form, the only difference being that the factor multiplying the steel parameter is 0·59 instead of 0·771. But M_u is not unduly sensitive to the value of this coefficient.

If the beam is over-reinforced, the following simplified procedure may be used. The Code permits the stress-block to be replaced by a uniform compression of $0.40f_{cu}$. Taking into account the shape of the stress-block (consisting of a parabolic arc and a rectangle), it is reasonable to assume that the centre of compression is located at $0.40nd$ from the top. Let the neutral axis depth = nd. Total compression = $0.40f_{cu}nbd$. Lever arm = $d(1-0.4n)$. The resisting moment

$$M_u = 0.40f_{cu}nbd^2(1-0.4n) \quad (4.12)$$

We also know that

$$j = (1 - 0.4n) \quad (4.13)$$

Knowing the steel parameter for a given beam, j may be obtained from (4.10). From (4.13), n is next found. It is now possible to insert this value of n in (4.12) and compute the moment of resistance M_u.

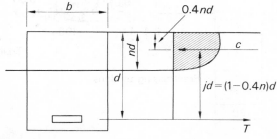

Fig. 4.10 Lever arm

38 Modern prestressed concrete design

The procedures outlined above for computing the ultimate flexural strength of under-reinforced and over-reinforced beams are also applicable to flanged sections when the neutral axis is within the flange.

Example 4.3

A pretenstoned beam shown in Fig. 4.11 is prestressed by 20 strands of 12·5 mm nominal diameter, the characteristic strength of each being 165 kN. The cube strength f_{cu} of the concrete is 50 N mm^2.

Check its safety at the ultimate limit state, given that the girder bending moment is 350 kN m and the live load bending moment is 450 kN m.

Solution

$$\left.\begin{array}{l}\text{Bending moment at the}\\ \text{ultimate limit state}\end{array}\right\} = 1\cdot40 \times 350 + 1\cdot60 \times 450 = 1200 \text{ kN m}$$

Assume a trial neutral axis of 16 cm from top (Fig. 4.11).
Total compression due to (1), (2), (3) and (4)

$$= \left(\frac{2250 \times 9\cdot6 \times 100}{1000}\right) + \left(\frac{1934 \times 2\cdot4 \times 100}{1000}\right)$$

$$+ \left(\frac{2}{3} \times \frac{2\cdot4 \times 316 \times 100}{1000}\right) + \left(\frac{2}{3} \times \frac{4 \times 30 \times 1934}{1000}\right)$$

$$= 2160 + 464 + 51 + 155 = 2830 \text{ kN}$$

Distance of centre of compression from top

$$= \left(\frac{2160 \times 4\cdot8 + 464 \times 10\cdot8 + 51 \times 10\cdot5 + 155 \times 13\cdot5}{2830}\right)$$

$$= 6\cdot36 \text{ cm}$$

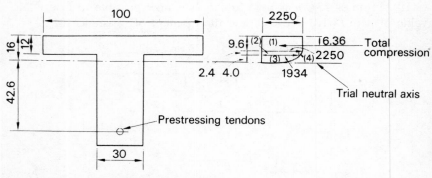

Fig. 4.11 Compression stress block

Strain in steel $= \dfrac{0\cdot0035}{16} \times 42\cdot6 +$ strain due to effective prestress

$= 0\cdot009\,31 + \dfrac{0\cdot8 \times 115\cdot5}{94\cdot2 \times 200} = 0\cdot009\,31 + 0\cdot0049 = 0\cdot014\,21$

As this strain exceeds the strain of $0\cdot0126$ corresponding to f_{pu}/γ_m found from Fig. 4.7, it may be assumed that the steel stress is equal to f_{pu}/γ_m. The total tension in 20 strands $=(165/1\cdot15)\times20 = 2869$ kN, which is very nearly equal to the total compression of 2830 kN. This implies that the assumed position of the neutral axis is fairly accurate. If more accuracy is desired, one more trial calculation may be made to ensure that they are exactly equal.

$M_u = 2830 \times (58\cdot6 - 6\cdot36)$

$= 1478\cdot39$ kN m $>$ B.M. of 1200 kN m.

Hence, the design is safe.

References

[4.1] Magnel, G., *Prestressed Concrete*, Concrete Publications, London, 3rd (revised and enlarged) edn, 1954.

[4.2] Ramaswamy, G. S., and Narayana, S. K., 'The ultimate flexural strength of post-tensioned rectangular beams', Third FIP Congress, Berlin, 1958.

5
Design for shear and torsion

Until very recently, design for shear was based on checking for principal stresses under working loads. Current code provisions in CP 110:1972, ACI 318:1971 and European Codes recognize that it is far more meaningful to check for shear and design shear reinforcement based on the behaviour of the prestressed concrete member at the ultimate limit state. This is because of the realization that stirrups and other shear reinforcement do not become active until the shear carrying capacity of the concrete is exhausted. If the ultimate shear force at a section is V_u and the shear that the concrete can carry is V_c, shear reinforcement needs to be designed to carry the residual shear force $V_u - V_c$.

5.1 Demarcation of Zones

Four different zones may be distinguished in a prestressed concrete beam (Fig. 5.1) for the purpose of designing shear reinforcement [5.1]. The charactereristics of the four zones, designated A to D, are as follows:

Zone A: In a simply supported beam, this zone extends to a distance d from the supports. This region remains crack-free up to failure.

Zone B: In this zone, the shear cracks develop within the web, when the principal tensile stress exceeds the tensile strength of the concrete. The cracks, following the principal stress (tension) trajectories are inclined at 20° to 30°. These cracks are generally referred to as *web-shear cracks*.

Zone C: In this zone of high bending moment, vertical flexural cracks are at first formed; with further increase in the load, they penetrate the beam and become inclined at 40° to 70°.

Zone D: In this zone where bending moment is dominant and shear is insignificant, only vertical flexural cracks are formed and shear failure can be ruled out.

Design for shear and torsion 41

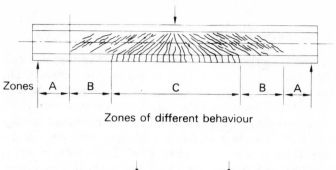

Zones of different behaviour

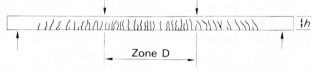

Zone D

$M/Th > 7$ no danger of shear failure T = shear

Rectangular beam: ratio $M/Th > 7$

Fig. 5.1 (After Leonhardt vide reference [5.1])

It is convenient to merge regions A and B and regard the combined regions as the *zone of web-shear cracking*. Region C, the zone of *flexural-shear cracking*, needs separate treatment. In CP 110, the shear-carrying capacity of the concrete in the region A–B is designated as V_{co} and the shear carrying capacity of concrete in Region C is designated as V_{cr}. Methods given in the Code for computing V_{co} and V_{cr} will now be explained.

5.2 Expression for V_{co}

In this zone, V_c designated as V_{co}, corresponds to the shear at which the principal tensile stress at the centroid of the section just attains a value of $f_t = 0.24\sqrt{f_{cu}}$. In computing the principal stress, the compressive stress at the centroid of the section is taken as $0.80f_{cp}$, where f_{cp} (regarded as positive) is the stress at the centroid due to prestressing. By applying the formula for principal stresses, we have

$$-f_t = \frac{0.80f_{cp}}{2} - \sqrt{(0.40f_{cp})^2 + \left(\frac{3}{2}\frac{V_{co}}{bh}\right)^2}$$

Or,

$$\tfrac{2}{3}\sqrt{f_t^2 + 0.8f_{cp} \cdot f_t} = \frac{V_{co}}{bh}$$

Hence
$$V_{co} = 0.67bh\sqrt{f_t^2 + 0.80f_{cp} \cdot f_t} \tag{5.1}$$

For I, T and L sections, $Ib/A\bar{y}$ at the centroid section will be greater than $0.67bh$ and the above expression will be conservative. The point of maximum principal stress for such sections will not, however, be the centroid but at the junction of the flange and the web, and checking at the centroid according to the simplified procedure would be slightly on the unsafe side. The two effects mentioned are compensating and hence formula (5.1) may be applied in all cases without appreciable error. If inclined tendons are provided, the vertical component of the force carried by them may be added to V_{co} when that component is such as to reduce the shear force at the section; if, on the other hand, it adds to the imposed shear, it is to be subtracted.

5.3 Expression for V_{cr}

In this zone V_c, designated as V_{cr}, is computed as

$$V_{cr} = \left(1 - 0.55\frac{f_{pe}}{f_{pu}}\right)v_c bd + M_0 \frac{V}{M} \tag{5.2}$$

where

f_{pe} = effective prestress in the tendon after losses. For this purpose, f_{pe} shall not be greater than $0.60f_{pu}$. v_c = ultimate shear stress in beams, and may be obtained from Table 5 of the Code.

f_{pu} = characteristic strength of the prestressing tendons. V and M are the shear and bending moment at the section at ultimate loads.

M_0 = the moment required to produce zero stress in concrete at a depth d from the extreme compression concrete fibre i.e., $M_0 = 0.80f_{pt}(I/y)$ where f_{pt} is the stress due to prestress only at depth d, y is the distance from the centroid to the point at a depth d from the extreme compression fibre, and I is the moment of inertia of the section about the centroidal axis.

The value of V_{cr} obtained from (5.2) shall not be less than $0.10bd\sqrt{f_{cu}}$.

The formula (5.2) is derived as follows. Experiments have established that

$$V_{cr} = 0.037bd\sqrt{f_{cu}} + \frac{M_c}{\left(\dfrac{M}{V} - \dfrac{d}{2}\right)} \tag{5.3}$$

This is also the form in which V_{cr} is given in ACI 318:1971. A close analysis of (5.3) shows that its first term accounts for the tensile strength of the concrete and the second term represents the shear on the onset of cracking under the cracking moment M_c. The occurrence of $d/2$ in the denominator accounts for the additional shear necessary to transform the flexural crack formed at a distance of $d/2$ from the section under consideration into an inclined shear crack at the section. Ignoring this effect, which will only result in a conservative estimate of V_{cr}, we may write

$$V_{cr} = 0{\cdot}037 bd\sqrt{f_{cu}} + M_c \frac{V}{M} \qquad (5.4)$$

The cracking moment M_c may be written as

$$M_c = M_0 + 0{\cdot}37\sqrt{f_{cu}}\,\frac{I}{y} \qquad (5.5)$$

where,

$0{\cdot}37\sqrt{f_{cu}}$ is the tensile strength of the concrete.

Substituting (5.5) in (5.4),

$$V_{cr} = 0{\cdot}037 bd\sqrt{f_{cu}} + 0{\cdot}37\sqrt{f_{cu}}\,\frac{I}{y}\frac{V}{M} + M_0\frac{V}{M} \qquad (5.6)$$

Noting that I/y for a rectangular section is $bd^2/6$ and that at a section where shear-flexure cracks occur $M \approx 4Vh$,

$$\frac{V_{cr}}{bd} = 0{\cdot}037\sqrt{f_{cu}} + \frac{d}{h}\frac{0{\cdot}37\sqrt{f_{cu}}}{6\times 4} + \frac{M_0 V}{Mbd} \qquad (5.7)$$

It is conservative to assume $d = h$. For a concrete with $f_{cu} = 50$ which is quite representative,

$$\frac{V_{cr}}{bd} = 0{\cdot}26 + 0{\cdot}11 + \frac{M_0 V}{Mbd} \qquad (5.8)$$

Equation (5.3) with which we started is based on experiments carried out on prestressed beams with a high degree of prestress of over $0{\cdot}50 f_{pu}$. In CP 110, the effort has been to develop a single formula for V_{cr} applicable to beams with different degrees of prestress so that Class 2 and Class 3 beams, representing the transition from fully prestressed concrete to reinforced concrete can also be covered. This may be done by rewriting the formula for V_{cr} in such a manner that it reduces, in the limit, to the experimentally established formula for beams with $f_{pe}/f_{pu} = 0{\cdot}60$ at one end and the formula for reinforced concrete beams at the other corresponding to $f_{pe}/f_{pu} = 0$. This may be

44 Modern prestressed concret design

done by writing the expression for V_{cr} as

$$V_{cr} = \left(1 - n\frac{f_{pe}}{f_{pu}}\right)v_c bd + M_0 \frac{V}{M} \tag{5.9}$$

For $\frac{f_{pe}}{f_{pu}} = 0.60$, equation (5.9) must reduce to

$$\frac{V_{cr}}{bd} = 0.37 + \frac{M_0}{bd}\frac{V}{M}$$

Or,

$$(1 - 0.6n)v_c = 0.37 \tag{5.10}$$

Hence, $n = 0.55$, assuming $v_c = 0.55$ for 0.50% of steel. Or,

$$V_{cr} = \left(1 - 0.55\frac{f_{pe}}{f_{pu}}\right)v_c bd + \frac{M_0 V}{M} \tag{5.11}$$

5.4 Design of Shear Reinforcement

If the shear force V at a section is less than V_c (the lesser of the two values V_{co} and V_{cr}) only, nominal reinforcement is necessary. If $V < 0.50 V_c$, there is no need for any reinforcement. Such is the case in slabs, pile caps and walls.

It needs to be checked first of all if the shear carried by the beam is less than the maximum permissible value given in Table 6 of the Code. This limitation is necessary to ensure that the web is not crushed by diagonal compression caused by excessive shear.

It has already been explained that the function of inclined stirrups, vertical stirrups or bent-up bars is to carry the residual shear $(V_u - V_c)$ at those sections where V_u exceeds V_c (i.e., the lesser of the two values V_{co} and V_{cr} at that section). For this purpose, it is necessary to find the shear carrying capacity of different types of web reinforcement. Expressions for shear carrying capacity of inclined stirrups and bent-up bars may be derived by using the classical truss analogy due to Mörch. In this analogy, it is assumed that the tensile forces are carried by the longitudinal and shear reinforcement in the form of bent-up bars or stirrups, and that the thrusts in the compression zone and the web are carried by the concrete.

In Figs. 5.2(a) and 5.2(b), the analogy is presented for inclined stirrups and bent-up bars. Let the cutting plane A-A intersect n inclined stirrups. Assuming that the residual shear $(V - V_c)$ is carried by the truss, the equation of equilibrium for forces in the vertical

Design for shear and torsion 45

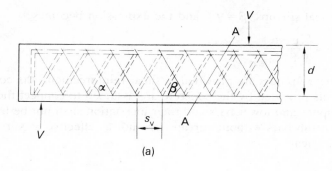

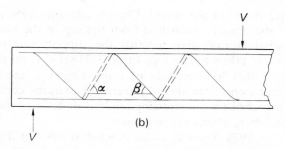

Fig. 5.2 (Truss systems for shear (a) with stirrups (b) with bent-up bars reproduced from *Handbook on the Unified Code for Structural Concrete*)

direction leads to

$$(v - v_c)bd = nA_{sv}(0\cdot 87 f_{yv}) \sin \beta \qquad (5.12)$$

where

$$n = \frac{d(\cot \alpha + \cot \beta)}{s_v}$$

Experimental investigations have shown that the inclined compression struts may be assumed at 45° to the longitudinal axis of the beam. Making this assumption,

$$(v - v_c)bd = 0\cdot 87 f_{yv} \cdot A_{sv}(\sin \beta + \cos \beta) \frac{d}{s_v} \qquad (5.13)$$

The area of shear reinforcement

$$\frac{A_{sv}}{s_v} = \frac{b(v - v_c)}{0\cdot 87 f_{yv}(\cos \beta + \sin \beta)} \qquad (5.14)$$

46 Modern prestressed concrete design

For vertical stirrups, $\beta = 90°$ and the expression becomes,

$$\frac{A_{sv}}{s_v} = \frac{b(v - v_c)}{0 \cdot 87 f_{yv}} \tag{5.15}$$

When bent-up bars are provided as shear reinforcement, the compression struts provided by the concrete are assumed to connect the bends in the upper and lower bars and their inclination shall not be less than 45°. Bent-up bars without stirrups, are not as effective as stirrups for carrying shear.

5.5 Comments on BS Code Clauses Relating to Shear

(1) The effective depth in the British Code is defined as the depth to the centroid of the tendons measured from the top of the beam. This provision is rather conservative towards the support section where several tendons are taken to the top. In ACI 318:1971, on the other hand, it is stated that for purposes of computing V_{co}, the effective depth d shall be taken as the distance from the extreme compression fibre to the centroid of the prestressing tendons or 80% of the overall depth of the member, whichever is greater.

(2) In ACI 318:1971, there is a clause which states that if a section at a distance $d/2$ from the support is closer than the transmission length of wire or strand used, a reduced prestress in concrete at sections falling within the transmission length shall be considered in computing V_{co}. The prestress at the centroid of the section shall be assumed to vary linearly from zero at the end of the member to its full value at a distance equal to the transmission length from the end of the member. The BS Code has no corresponding provision.

(3) The object of writing the first term in the form given (5.11) is ostensibly to make it applicable to all categories of beams ranging from Class 1 to Class 4. However, a close examination of the first term of (5.11) will show that the ratio f_{pe}/f_{pu} can only take into account the degree to which a tendon is prestressed. It is not an index of the degree of prestress in the beam. If the degree of prestress in the beam is to be introduced to make the formula applicable to Class 2 and Class 3 beams, the relative contributions of the prestressing tendons and the untensioned steel will have to be inserted into (5.11) in some form. The Swiss Code SIA 162:1968 [5.2] attempts to do this in the manner described in the next article.

5.6 Provisions of Swiss Code SIA 162:1968

The Code states that the different shear resistances at the section added together must be greater than the ultimate shear V_u at the

Design for shear and torsion

section. This leads to the inequality

$$V_u \leq V_c + V_N + V_B + V_D \tag{5.16}$$

where, V_c, the shear carried by the concrete compression zone is given by

$$V_c = \left(1 + \frac{P_\infty}{Z_s}\right) v_c bd \leq 1{\cdot}5 v_c bd \tag{5.17}$$

P_∞ = Prestressing force at the section after losses
Z_s = The total tensile force in the tendons (prestressed and non-prestressed) in the tension zone at the time of yielding.
$V_N = 0{\cdot}2 f_{cp} bd = 0{\cdot}20 P_\infty$, where, f_{cp} is the prestress at the centre of gravity of the cross-section calculated from the prestressing force. V_N is permitted to be taken into account only in Zone A–B where the web is uncracked by flexure.

V_B and V_D are the shear-carrying capacities of inclined and vertical stirrups to be computed by formula (5.13). For the purposes of designing shear reinforcement, it may perhaps be better to recast (5.16) as

$$V_B + V_D \geq V_u - V_c - V_N \tag{5.18}$$

Shear reinforcement, in accordance with (5.18), is to be provided if the shear stress v_u at a section before the ultimate limit state exceeds v_c. Even otherwise, nominal reinforcement is to be provided such that

$$V_B + V_D = \tfrac{1}{2} v_c bd \tag{5.19}$$

The Swiss Code also limits the maximum shear stress and the maximum stirrup spacing.

5.7 Experimental Verification

A series of tests on prestressed beams with different degrees of prestress are now in progress at the Structural Engineering Research Centre, Roorkee, India to assess the merits of the procedures for the design of shear reinforcement recommended in the BS and Swiss Codes.

The details relating to one of the test beams in the series shown in Fig. 5.3 are summarized below. The British and Swiss Code provisions will be applied to the test beam and compared with the experimental results.

Data

$b = 10$ cm; $d = 42$ cm
$I = 120{,}792$ cm^4
$A_c = 615$ cm^2

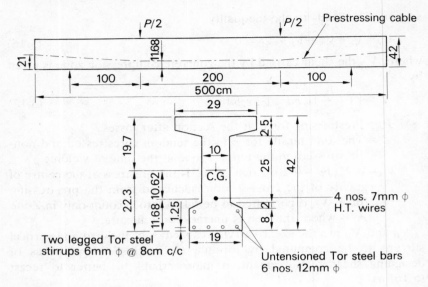

Fig. 5.3 Details of test beam

Area of prestressing steel	$= 1 \cdot 52 \text{ cm}^2$
Area of non-prestressed steel	$= 6 \cdot 78 \text{ cm}^2$
Ultimate strength of prestressing steel	$= 1575 \text{ N/mm}^2$
0·2% Proof stress of prestressing steel	$= 1385 \text{ N/mm}^2$
Ultimate strength of non-prestressed deformed bars	$= 608 \cdot 6 \text{ N/mm}^2$
0·2% Proof stress of non-prestressed deformed bars	$= 493 \cdot 0 \text{ N/mm}^2$
Average cube strength of concrete	$= 41 \cdot 7 \text{ N/mm}^2$
Initial prestressing force	$= 18 \cdot 0 \text{ t}$
Final (effective) prestressing force	$= 15 \cdot 3 \text{ t}$

Stirrups provided: 6 mm diameter two-legged stirrups at 8 cm centre to centre

Experimental Data

Flexural Cracking Load P	$= 9 \cdot 0 \text{ t}$
Load P corresponding to first diagonal crack (formed as an extension of the first flexural crack)	$= 14 \cdot 0 \text{ t}$
Load P at ultimate	$= 39 \cdot 0 \text{ t}$

Design for shear and torsion 49

Calculation of v_{cr} according to CP 110:1972
The section is cracked in flexure. Hence

$$V_{cr} = \left(1 - 0.55 \frac{f_{pe}}{f_{pu}}\right) v_c bd + M_0 \frac{V}{M}$$

$$\frac{f_{pe}}{f_{pu}} = \frac{153\,000}{1575 \times 152} = 0.65$$

$0.65 > 0.60$ and in accordance with the Code f_{pe}/f_{pu} will be limited to 0.60.

A_s = total area of longitudinal steel
  $= 1.52 + 6.78 = 8.30 \text{ cm}^2$

$$\frac{A_s}{bd} \times 100 = \frac{8.30 \times 100}{42 \times 10} = 1.97.$$

v_c may now be read off from Table 5 of CP 110 as 0.95 N/mm^2.

$$\frac{V}{M} = \frac{V}{Va} = \frac{1}{a} = \frac{1}{1000} \text{ mm}^{-1} = 1.0 \text{ m}^{-1}$$

$$f_{pt} = \frac{153\,000}{61\,500} + \frac{153\,000 \times 106.2 \times 106.2}{120\,792 \times 10^4} = 3.9 \text{ N/mm}^2$$

$$M_0 = 0.80 \times 3.9 \times \frac{120\,792 \times 10^4}{106.2} \times \frac{1}{10^3} = 35\,490 \text{ Nm}$$

$$V_{cr} = (1 - 0.55 \times 0.60) \times 0.95 \times 100 \times 420$$
$$+ 35\,490 = 62\,190 \text{ N} \approx 6.2 \text{ t}$$

$$V_{cr} \text{ observed in the test} = \frac{P}{2} = \frac{14}{2} = 7.0 \text{ t}$$

Hence, the agreement is good.

Shear resistance contributed by stirrups (CP 110)
  A_{sv} = area of two-legged stirrups = 0.566 cm^2
  s_v = spacing of stirrups = 8 cm
  d = effective depth = $42.0 - 1.25 = 40.75$ cm
  f_{yv} = 0.20% proof stress = 493 N/mm^2
But is is restricted to 425 N/mm^2 in accordance with Code provisions.

  Shear resistance of stirrups = $0.87 f_{yv} \cdot A_{sv} \cdot \dfrac{d}{s_v}$

$$= \frac{0.87 \times 425 \times 407.5 \times 56.6}{80} = 106\,600 \text{ N} = 10.66 \text{ t}$$

Hence total shear resistance of the beam including stirrups = $V_{cr} + 10\cdot 66 = 6\cdot 20 + 10\cdot 66 = 16\cdot 86$ t. This compares fairly well with the failure load $\dfrac{P}{2} = \dfrac{39}{2} = 19\cdot 5\, t$ observed in the experiment.

Shear calculations according to Swiss Code SIA 162:1968

Shear resistance of stirrups

$$= \frac{A_{sv} \cdot f_{yv} d}{s_v}$$

$$= \frac{56\cdot 6 \times 493\cdot 0 \times 407\cdot 5}{80}$$

$$= 142\,500 \text{ N}$$

$$= 14\cdot 25 \text{ t}$$

Shear carried by compression zone in accordance with formula (5.17)

$$= \left(1 + \frac{f_{pe} A_{ps}}{f_{py} A_{ps} + f_y A_s}\right) v_c bd$$

$$\leqslant 1\cdot 5 v_c bd$$

$f_{pe} \cdot A_{ps} = 15\cdot 3$ t

$f_{py} A_{ps} + f_y A_s = 138\cdot 5 \times 1\cdot 52 \times 100 + 49\cdot 3 \times 6\cdot 70 \times 100$

$$= 54\,400 \text{ kg} = 54\cdot 4 \text{ t}$$

$v_c = 12$ kg/cm^2 according to SIA 162:1968

$$V_c = \left(1 + \frac{15\cdot 3}{54\cdot 4}\right) \times 12 \times 40\cdot 75 \times 10 = 6270 \text{ kg}$$

$$= 6\cdot 27 \text{ t}$$

This agrees closely with that given by CP 110.
Hence total shear resistance of beam = $14\cdot 25 + 6\cdot 27 = 20\cdot 52$ t against the experimentally observed value of the ultimate shear = $\frac{39}{2} = 19\cdot 5$ t.

Note: It will be observed that both the Swiss Code and CP 110 give more or less the same results for the shear carried by the concrete. The shear carried by stirrups according to CP 110 is less for two reasons. Firstly, the proof stress is limited to 425 N/mm^2 by the Code; secondly, $\gamma_m = 1\cdot 15$ has been introduced which in turn introduces a factor of 0·87 which is not present in the Swiss Code provisions.

5.8 Torsion in Prestressed Concrete

Current thinking on torsion in prestressed concrete members forms the basis of CEB-FIP recommendations [5.3] and is clearly explained by Leonhardt [5.1] and Lampert [5.4].

A clear distinction is made between compatibility and equilibrium torsion. The former is the result of restraint against twist in a statically indeterminate structure. The latter is required to maintain equilibrium in a statically determinate structure. Experiments have shown that on the onset of cracking there is a steep drop in torsional stiffness amounting to as much as 4 to 8 times the decrease in flexure stiffness. Consequently, the compatibility torsion soon becomes negligible and may be ignored. On the other hand, members subject to equilibrium torsion need to be adequately designed to carry it without distress.

Lampert [5.4] develops a space-truss model for members subject to pure torsion. The longitudinal reinforcement concentrated at the corners constitutes the top and bottom booms of the truss; the closed stirrups form the verticals, and concrete provides the compression diagonal between cracks. Experiments have shown that after cracking only a thin outer shell of a solid rectangular section carries the load If the member is under-reinforced, there is no difference in the ultimate torque carried by a solid and a hollow section. Hence it is possible to replace solid sections by a hollow-section with a certain thickness t_1, the centre line of the hollow-section being defined by the lines joining the centres of the longitudinal bars provided at the centres. Sometimes these centre lines are approximated by the centre lines of the stirrups forming closed loops.

Let us first consider the basic case of a hollow rectangular section (Fig. 5.4) subjected to a torsional moment T due to ultimate loads. Let v_t be the shear stress induced. It is easily seen that

$$v_t h_0 b_0 t + v_t b_0 h_0 t = T \qquad (5.20)$$

Or, $v_t = \dfrac{T}{2 b_0 h_0 t}$

Denoting $b_0 h_0$ by A_0,

$$v_t = \dfrac{T}{2 A_0 t} \qquad (5.21)$$

It is necessary to restrict this shear stress in the concrete to permissible values prescribed in the Codes, e.g., Table 7 of CP 110 to prevent the buckling of the concrete compression diagonal. If the section is a solid rectangle $b \times d$ it may be reduced to a hollow rectangle by assuming a nominal thickness of $t_1 = b/6$ or $b_0/5$ whichever is less, where b and b_0

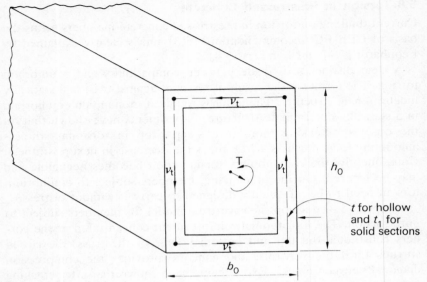

Fig. 5.4 Hollow or solid section under torsion

are the diameters of the largest inscribed circles in the given cross-section and the area A_0.

If the volumes of longitudinal steel and closed loop stirrups are equal, the space truss model will have 45° diagonals. We will also have the relationship

$$A_{sl}s_v f_{yl} = A_{sv}f_{yv}\frac{u}{2} \qquad (5.22)$$

where
$\quad\quad u$ = perimeter of the closed stirrups
$\quad\quad\quad = 2(x_1 + y_1) \approx 2(b_0 + h_0)$
$\quad A_{sl}$ = area of longitudinal steel
$\quad A_{sv}$ = area of the closed links at a section (area of two legs)
$\quad x_1$ and y_1 are the smaller and larger dimensions of the closed link stirrups
$\quad\quad s_v$ = spacing of the closed link stirrups
$\quad\quad f_{yl}$ = characteristic strength of longitudinal reinforcement
$\quad\quad f_{yv}$ = characteristic strength of link steel

Equation (5.22) will be readily recognized to be the same form as equation 12 of Article 3.3.7 of CP 110:1972.

Using the space truss model with 45° diagonals, it may be shown that

$$T = 2A_0\left(\frac{A_{sl}f_{yl}}{u}\right) \qquad (5.23)$$

Design for shear and torsion

Substituting from equation (5.22), it may also be seen that

$$T = A_0 \frac{A_{sv} f_{yv}}{s_v} \qquad (5.24)$$

For design purposes, formulae (5.23) and (5.24) need to be modified by replacing f_{yl} and f_{yv} by $\frac{f_{yl}}{\gamma_m}$ and $\frac{f_{yv}}{\gamma_m}$ respectively to give

$$\frac{A_{sv}}{s_v} = \frac{T}{b_0 h_0 (0 \cdot 87 f_{yv})}$$

$$\approx \frac{T}{x_1 y_1 (0 \cdot 87 f_{yv})} \qquad (5.25)$$

CP 110:1972 introduces an additional factor of 0·80 in the denominator of equation (5.25) to give

$$\frac{A_{sv}}{s_v} \leqslant \frac{T}{0 \cdot 80 x_1 y_1 (0 \cdot 87 f_{yv})} \qquad (5.26)$$

and

$$A_{sl} \geqslant \frac{A_{sv}}{s_v} \frac{f_{yv}}{f_{yl}} (x_1 + y_1) \qquad (5.27)$$

The above formulae permit the areas of longitudinal steel and closed link steel to be found.

The spacing of the links shall not exceed x_1, $y_1/2$ or 200 mm so that each crack is intersected by at least one link.

Where there is interaction between bending and torsion or bending, torsion and shear analysis may be carried out as indicated by Lampert [5.4].

References

[5.1] Leonhardt, F., 'Shear and torsion in prestressed concrete', Lecture delivered at Session IV, Sixth FIP Congress, Prague, June 1970.

[5.2] SIA Code 162, Edition 1968 *Norm für die Berechnung, Konstruktion und Ausführung von Bauwerken aus Beton, Stahlbeton and Spannbeton* (Code for the design, construction and execution of concrete, reinforced concrete and prestressed concrete structures) Schweizer Ingenieur und Architekten-Verein, Zurich (In German).

[5.3] *International Recommendations for the Design and Construction of Concrete Structures, Principles and Recommendations*, June 1970: Sixth FIP Congress, Prague (English edition), Section R.43.43.

[5.4] Lampert, P., 'Torsion and bending in reinforced concrete and prestressed concrete members', *Proceedings of the Institution of Civil Engineers*, December 1971, Vol. 50, pp. 487–505.

6
Design of pretensioned products

A large variety of standardized pretensioned roofing elements and bridge girders are being produced in the UK, USA and several European countries. In the USA, single tees, double tees, cored slabs and AASHO bridge girders have been standardized and catalogued [6.1]. In the UK, bridge girders have been standardized by the Prestressed Concrete Development Group [6.2]. Pretensioned hyperboloids, sometimes known as Silberkuhl shells, have been standardized for production in West Germany [6.3], India and Iraq [6.4].

In this chapter, the detailed design of two typical pretensioned products, namely, the double tee and the pretensioned hyperboloid will be illustrated by means of fully worked out examples.

6.1 Pretensioned Hyperboloids

Pretensioned hyperboloids offer a very economical, versatile and aesthetically attractive form of roofing for industrial buildings in the span range of 10 to 18 m. In the form in which the product is commonly used, it is a piece cut out of a hyperboloid of revolution of

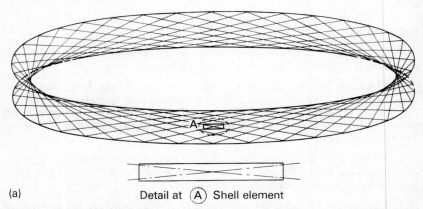

(a)　　　　　　　　　Detail at (A)　Shell element

Fig. 6.1(a)　Shell unit as part of a hyperboloid of revolution of one sheet

Design of pretensioned products 55

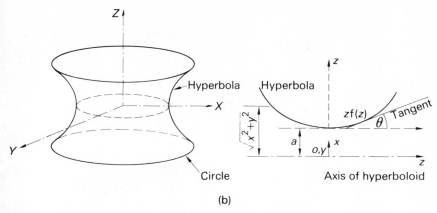

Fig. 6.1(b) Surface of the hyperboloid of revolution of one sheet

one sheet (Fig. 6.1(a)). It may be thought of as a translational surface generated by moving a hyperbola over an arc of a circle so that the cross-section is a hyperbola and the longitudinal section is an arc of a circle (Fig. 6.2). It is a doubly ruled surface offering the practical advantage that the prestressing tendons can be run straight to lie entirely on the surface. Its versatility as a roofing element is evident by looking at Fig. 6.3 where a number of roofing arrangements with and without northlights are shown. The easiest way to use them is to rest them side by side on two masonry walls with seating saddles cut in them to receive the shells (Fig. 6.4). The longitudinal curvature of the

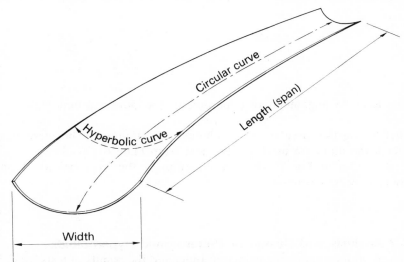

Fig. 6.2 Geometric form of the shell unit

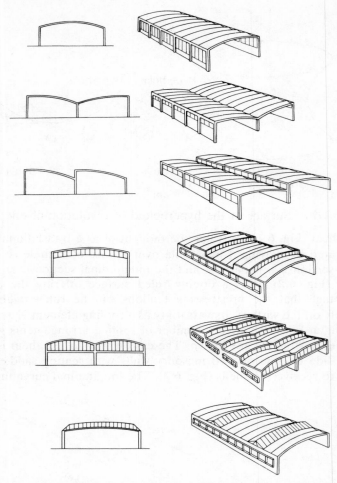

Fig. 6.3 Arrangements of hyperboloids for industrial buildings

shell along the circular arc facilitates easy drainage. The manner in which the units are used will be clear from Plate 6.1.

Referring to Fig. 6.1(b), the equation of the hyperboloid of one sheet may be written as

$$x^2 + y^2 - \frac{a^2}{c^2} z^2 = a^2 \tag{6.1}$$

6.2 Analysis and Design of Pretensioned Hyperboloids

Based on the experience gained in India and the results of tests [6.5], it has been established that it is sufficiently accurate to analyse and

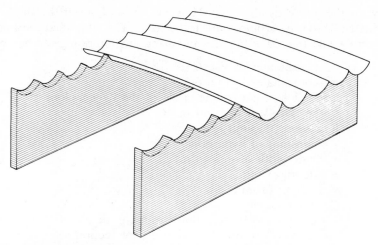

Fig. 6.4 Hyperboloids resting on masonry walls

construct pretensioned hyperboloids as beams of curved cross-section. Following the usual procedure for the design of pretensioned beams, four Magnel inequalities are set up, two of them being at the support at transfer and the other two at midspan under working loads. Using Magnel's graphical method already explained in Chapter 4, the pre-stressing force and its eccentricity are found. Checks for the ultimate limit states of flexure and shear are carried out in the same manner as for beams. However, tedious calculations are involved in computing the sectional properties of the hyperbolic cross-section. These computations can be drastically simplified if the hyperbolic curve is approximated by a parabola as suggested by Ramaswamy and Nabil Rafiq Saeed [6.5]. The hyperbolic cross-section is replaced by a parabola

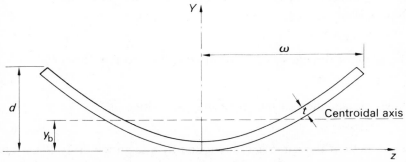

Fig. 6.5 Cross-section of hyperboloid with hyperbola approximated as parabola

only for the purposes of design. However, in setting out the shell, the ordinates of the hyperboloid are to be used. The simplified method is presented with the aid of Fig. 6.5 and using the notation given below

    x, y, z—coordinates in Fig. 6.1
    x, Y, z—coordinates in Fig. 6.5
        a = radius of generating circle of the hyperboloid
        t = thickness of the unit in m
        h = rise of the unit
        c = constant of hyperbola
        w = half width of unit in m
        d = depth of unit
        W = weight of unit
        C = crane capacity
        Y_b = distance to bottom fibre from c.g. of section
        Y_t = distance to top fibre from c.g. of section
        θ = slope of tangent to unit at its ends
        l = span of unit
        L = length of unit in m
        A_c = area of concrete cross-section
        k = constant of parabola
        γ = density of concrete in kg/m^3

The other notations used have the usual meanings.

6.3 Simplified Method of Analysis

A simplified method of analysis demanding only slide rule accuracy results if the hyperbolic curve is replaced by a parabolic arc (Fig. 6.5) with the equation

$$Y = kz^2 \tag{6.2}$$

Casting difficulties arise if the slope dY/dz exceeds 45° at $z = w$. Noting that,

$$dY/dz = 2kz \tag{6.3}$$

the slope at the ends $= 2kw$. Restricting $2kw$ to tan $45 = 1 \cdot 0$,

$$k = 1/2w. \tag{6.4}$$

Now, the equation of the curve may be written as

$$Y = z^2/2w \tag{6.5}$$

The length of the curve is easily found by integration as $2 \cdot 2956w$ and hence

$$A_c = 2 \cdot 2956wt \tag{6.6}$$

Design of pretensioned products

Similarly, it can be shown that

$$Y_b = 0.1830w \text{ and} \tag{6.7}$$
$$Y_t = 0.3170w \tag{6.8}$$

The moment of inertia and moduli of section are found to be

$$I_{zz} = 0.05376w^3 t \tag{6.9}$$
$$Z_b = 0.2937w^2 t \tag{6.10}$$
$$Z_t = 0.1696w^2 t \tag{6.11}$$

Experience gained in designing several hyperboloids indicates that of the four inequalities relating to stresses in fibres, the governing equality for the magnitude of the prestressing force is the condition of no tension or small residual compression in the bottom fibre at midspan under working loads. As no tension is permitted, the pressure line under M_T can shift only up to the top kern point.
Hence

$$M_T = \left(\frac{Z_b}{A_c} + e\right) P_e$$

Or

$$P_e = \frac{M_T}{\left(\dfrac{Z_b}{A_c} + e\right)} \tag{6.12}$$

It is advantageous to make e as large as possible. Allowing for a cover of 3 cm, the maximum practicable eccentricity is

$$e = Y_b - 3 \tag{6.13}$$

where e and Y_b are in cm. At the ends, the bending moment being zero, it is desirable to have the centre of prestress coincide with the centre of gravity of the concrete section. Simple geometrical considerations will show that this condition will be realized if

$$h = e \tag{6.14}$$

where h is the rise of the shell. The rise of the shell is linked to a, the radius of circular arc by the relationship

$$e = h \approx \frac{l^2}{8a}$$

Hence the appropriate radius

$$a = \frac{L^2}{8e} \tag{6.15}$$

60 Modern prestressed concrete design

Based on this approximate analysis, the following step-by-step procedure may be followed.

Step 1
Ascertain the available handling capacity C in tons which decides the weight W of the shell unit so that

$$W \leq C \qquad (6.16)$$

$$W = C = \frac{2 \cdot 2956 L \gamma w t}{1000}$$

Knowing L and assuming $t = 0 \cdot 06$ m; w, the semi-width of the shell is found.

Step 2
Knowing w, Y_b, it is possible to calculate A_c and e.

Step 3
P_e is calculated from (6.12) and P is found from the relation $P = 1 \cdot 25 P_e$, the losses in prestress being assumed as 20% of the initial prestress.

Step 4
The radius a of the shell is found from (6.15).

Step 5
Stresses at transfer and working loads are checked.

Step 6
The ultimate moment of resistance is checked to see if it is adequate to equal or exceed the external bending moment.

Step 7
The shear stresses are checked.

This step-by-step procedure is illustrated in Design Example 6.2.

Design Example 6.1
Design of a pretensioned roof member: Class 1 design.

Data
    Span = 20 m c to c of supports
    Live load = 175 kg/m^2

Table 6.1 I about top

Designation	Area	I about own centroid	y	y^2	Ay^2	I
I	960	1 280	2	4	3 840	5 120
II	615	1 230	6	36	22 140	23 370
III	210	630	7	49	10 290	10 920
IV	56·26	176	12·5	156	8 780	8 956
V	1 000	208 333	35	1 225	1 225 000	1 433 333

Total $I_t = 1\ 481\ 699\ \text{cm}^4$

Materials

Concrete $f_{cu} = 50\ \text{N/mm}^2$
$f_{ci} = 40\ \text{N/mm}^2$

Steel

7 wire strands of 12·5 mm diameter. Nominal cross-sectional area = 94·2 mm².
Characteristic strength $A_{ps}f_{pu} = 165\ \text{kN}$.

Allowable stresses

Compressive stress in concrete at transfer $\Big\} = 0·50f_{ci} = 20\ \text{N/mm}^2$
Compressive stress at serviceability limit state $\Big\} = 0·33f_{cu} = 16·67\ \text{N/mm}^2$
Tensile stresses at transfer $= 1·00\ \text{N/mm}^2$
Losses in prestress $= 20\%$ of initial prestress
Hence $K = 0·80$

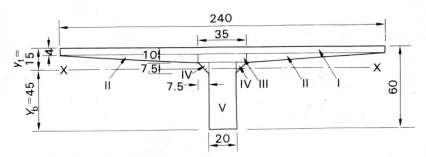

Fig. 6.6 Section of 'T' beam

62 Modern prestressed concrete design

Assume depth of 60 cm and adopt a section as sketched in Fig. 6.6.

$$y_t = \frac{1920 + 1845 + 1845 + 1470 + 35\,000 + 700}{2841}$$

$$y_t = \frac{42\,780}{2841} = 15\cdot 0 \text{ cm}$$

$I_{xx} = 1\,481\,699 - 2841 \times 225 = 842\,474 \text{ cm}^4$

$y_t = 15\cdot 0$

$y_b = 45\cdot 0$

$$Z_b = \frac{842\,474}{45} = 18\,722 \text{ cm}^3$$

$$Z_t = \frac{842\,474}{15} = 56\,164 \text{ cm}^3$$

$$\text{Self-wt/m} = \frac{2841}{10^4} \times 23 \times 10^2 \times 10 \text{ N/m} = 6550 \text{ N/m}$$

$$M_G = \frac{6550 \times 20 \times 20}{8} = 327\cdot 5 \text{ kN m.}$$

$$\text{Live load moment} = \frac{175 \times 2\cdot 4 \times 20 \times 20 \times 10}{8}$$

$M_L = 210 \text{ kN m}$

$M_T = M_G + M_L = 537\cdot 5 \text{ kN m}$

Formulation of Magnel equations

At transfer

(1) $\dfrac{P}{A} + \dfrac{Pe}{Z_b} - \dfrac{M_G}{Z_b} \leq 2000$

$P\left(\dfrac{1}{A} + \dfrac{e}{Z_b}\right) \leq 2000 + \dfrac{M_G}{Z_b}$

$P\left(\dfrac{Z_b + Ae}{A}\right) \leq 2000 Z_b + M_G$

$P \leq \dfrac{(2000 Z_b + M_G) A}{Z_b + Ae}$

Design of pretensioned products

$$\frac{1}{P} \geq \frac{e}{(2000Z_b + M_G)} + \frac{Z_b}{A(2000Z_b + M_G)}$$

$$\frac{1}{P} \geq \frac{e}{(2000 \times 18\,700 + 32\,750\,000)} + \frac{18\,700}{2841 \times 10^7}\left(\frac{1}{1 \cdot 87 \times 2 + 3 \cdot 27}\right)$$

$$\frac{1}{P} \times 10^7 \geq \frac{e}{(3 \cdot 74 + 3 \cdot 275)} + \frac{18\,700}{2841 \times 7}$$

$$\frac{1}{P} \times 10^7 \geq \frac{e}{(3 \cdot 74 + 3 \cdot 275)} + \frac{18\,700}{2841 \times 7}$$

$$\geq \frac{e}{7 \cdot 0} + \frac{18\,700}{19\,887}$$

$$\frac{1}{P} \times 10^7 \geq 0 \cdot 143e + 0 \cdot 94 \qquad (6.17)$$

At $\dfrac{1}{P} = 0$, $e = -\dfrac{0 \cdot 94}{0 \cdot 143} = -6 \cdot 58$

At $e = 0$, $\dfrac{1}{P} \times 10^7 = 0 \cdot 94$

(2) $-\dfrac{P}{A} + \dfrac{Pe}{Z_t} - \dfrac{M_G}{Z_t} \leq 100$

$P\left(\dfrac{e}{Z_t} - \dfrac{1}{A}\right) \leq 100 + \dfrac{M_G}{Z_t}$

$P\left(\dfrac{Ae - Z_t}{A}\right) \leq 100Z_t + M_G$

$P \qquad \leq \dfrac{A(100Z_t + M_G)}{Ae - Z_t}$

$\dfrac{1}{P} \qquad \geq \dfrac{e}{(100Z_t + M_G)} - \dfrac{Z_t}{A(100Z_t + M_G)}$

$100Z_t + M_G = 100 \times 56\,100 + 32\,750\,000$
$\qquad\qquad = (0 \cdot 56 + 3 \cdot 27) \times 10^7$

$\dfrac{1}{P} \times 10^7 \geq \dfrac{e}{3 \cdot 83} - \dfrac{Z_t}{A \times 3 \cdot 83}$

$\qquad \geq \dfrac{e}{3 \cdot 83} - \dfrac{56\,100}{2841 \times 3 \cdot 83}$

$$\frac{1}{P} \times 10^7 \geq 0\cdot 26e - 5\cdot 16$$

When $P = 0$, $e = \dfrac{5\cdot 16}{0\cdot 26} = 19\cdot 85$

$$\frac{1}{P} \times 10^7 \geq 0\cdot 26e - 5\cdot 16 \qquad (6.18)$$

(3) $K\dfrac{P}{A} - K\dfrac{Pe}{Z_t} + \dfrac{M_t}{Z_t} \leq 1667$

$KP\left(\dfrac{1}{A} - \dfrac{e}{Z_t}\right) \leq 1667 - \dfrac{M_t}{Z_t}$

$KP\left(\dfrac{Z_t - Ae}{A}\right) \leq (1667 Z_t - M_t)$

$KP \leq \dfrac{(1667 Z_t - M_t)A}{Z_t - Ae}$

$\dfrac{1}{KP} \geq \dfrac{Z_t}{A(1667 Z_t - M_t)} = \dfrac{e}{(1667 Z_t - M_t)}$

$1667 Z_t - M_t = 1667 \times 56\,100 - 53\,750\,000$

$\qquad = (1\cdot 667 \times 5\cdot 6 - 5\cdot 375) \times 10^7$

$\qquad = (9\cdot 35 - 5\cdot 375) \times 10^7$

$\qquad = 3\cdot 98 \times 10^7$

$\dfrac{Z_t}{A} = \dfrac{56\,100}{2841} = 19\cdot 75$

Losses $\quad = 20\%$

$K \qquad\quad = 0\cdot 80$

$\dfrac{1}{P} \times 10^7 \geq \dfrac{15\cdot 80}{3\cdot 98} - \dfrac{0\cdot 80 e}{3\cdot 98}$

$\dfrac{1}{P} \times 10^7 \geq 3\cdot 97 - 0\cdot 20 e \qquad (6.19)$

when $\dfrac{1}{P} = 0$, $e = 19\cdot 85$

(4) $\dfrac{KP}{A} + \dfrac{KPe}{Z_b} - \dfrac{M_t}{Z_b} \geq 0$

$KP\left(\dfrac{1}{A} + \dfrac{e}{Z_b}\right) \geq \dfrac{M_t}{Z_b}$

$$KP\left(\frac{Z_b + Ae}{A}\right) \geqslant M_t$$

$$KP \geqslant \frac{M_t A}{Z_b + Ae}$$

Or, the last inequality may be written as

$$\frac{1}{P} \times 10^7 \leqslant \left(\frac{0\cdot80e}{5\cdot375}\right) + \left(\frac{0\cdot80 \times 18\,700}{2841 \times 5\cdot373}\right)$$

i.e., $\frac{1}{P} \times 10^7 \leqslant 0\cdot149e + 0\cdot98$ \hfill (6.20)

Value of $e = -6\cdot58$,
when $\frac{1}{P} = 0$

Summary of Magnel inequalities

$$\frac{1}{P} \times 10^7 \geqslant 0\cdot143e + 0\cdot94 \hfill (6.17)$$

$$\frac{1}{P} \times 10^7 \geqslant 0\cdot26e - 5\cdot16 \hfill (6.18)$$

$$\frac{1}{P} \times 10^7 \geqslant 3\cdot97 - 0\cdot20e \hfill (6.19)$$

$$\frac{1}{P} \times 10^7 \leqslant 0\cdot149e + 0\cdot98 \hfill (6.20)$$

The inequalities are shown plotted in Fig. 6.7.

Calculation of P from inequalities

The maximum e physically possible is 40 cm, allowing for a cover of 5 cm. We may write the inequalities for $e = 40$ cm as follows.
From (6.17)

$$\frac{1}{P} \times 10^7 > 5\cdot72 + 0\cdot94 > 6\cdot6$$

$$P \leqslant \frac{10^7}{6\cdot6} N \leqslant \frac{10^4}{6\cdot6} \leqslant 1515 \text{ kN}$$

The third constraint (6.19) is meaningful only so long as $3\cdot97 - 0\cdot20e$ is positive. This expression becomes zero at $e = 19\cdot85$. Thereafter, this

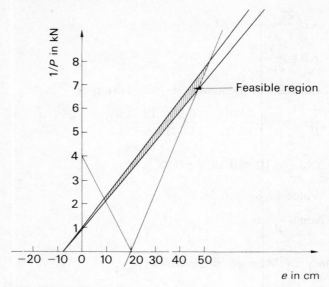

Fig. 6.7 Magnel diagram

constraint will cease to be active. Since the eccentricity considered is $40 \text{ cm} > 19 \cdot 85$, this constraint need not be considered.

From (6.18)

$$\frac{1}{P} \times 10^7 \geqslant 0 \cdot 26e - 5 \cdot 16, \quad e = 40$$

$$\geqslant 10 \cdot 4 - 5 \cdot 16 \geqslant 5 \cdot 24$$

$$P \leqslant \frac{10^4}{5 \cdot 24} \text{ kN} \leqslant 1908 \text{ kN}$$

From (6.20)

$$\frac{1}{P} \times 10^7 \leqslant 5 \cdot 96 + 0 \cdot 98$$

$$P \geqslant \frac{10^4}{6 \cdot 94} \geqslant 1441 \text{ kN}$$

Considering all the constraints, a prestressing force $P = 1500$ kN applied at an eccentricity of $e = 40$ cm appears suitable, provided there is provision for the strands with adequate cover.

Design of pretensioned products 67

To check if e = 40 cm is appropriate
The maximum eccentricity possible below the bottom kern point =
$$\frac{M_G}{P} = \frac{32\,750}{1500} = 21 \cdot 8 \text{ cm}$$

$$r^2 = \frac{I}{A} = \frac{841\,622}{2841} = 297 \text{ cm}^2$$

$$\frac{r^2}{y_t} = k_b = \frac{297}{15} = 19 \cdot 75 \text{ cm}$$

Hence, maximum eccentricity possible with no tension in top fibre $\Big\} = 21 \cdot 8 + 19 \cdot 75 = 41 \cdot 55 \text{ cm}$

Hence, eccentricity of 40 cm is satisfactory.

Check from Magnel diagram

$$\frac{1}{P} \times 10^7 = 6 \cdot 5 \quad \text{for} \quad e = 40 \text{ cm}$$

$$P = \frac{10^7}{6 \cdot 5} = 1540 \text{ kN}$$

But the graph has been drawn rather roughly.

Check on stresses at midspan

(1) At transfer $\qquad P = 1500 \text{ kN}; \quad e = 40 \text{ cm}$

(a) At bottom fibre

$$\frac{1\,500\,000}{2841} + \frac{1\,500\,000 \times 40}{18\,700} - \frac{32\,750\,000}{18\,700}$$

$$= 528 \quad + 3200 \quad\quad -1750 = 1978$$

$1978 < 2000 \text{ N/cm}^2$ which is satisfactory.

(b) At top fibre

$$-\frac{1\,500\,000}{2841} + \frac{1\,500\,000 \times 40}{56\,100} - \frac{32\,750\,000}{56\,100}$$

$$-528 \quad\quad +1067 \quad\quad -583 = 44$$

$44 < 100 \text{ N/cm}^2$ which is satisfactory.

68 Modern prestressed concrete design

(2) At service loads

(a) Bottom fibre

$$0{\cdot}80\times\left(\frac{1\,500\,000}{2841}+\frac{1\,500\,000\times 40}{18\,700}\right)-\frac{53\,750\,000}{18\,700}$$

$0{\cdot}80\times(528+3200)-2888 = 94$ N/cm^2, which is satisfactory.

(b) At top fibre

$$0{\cdot}80\times\left(\frac{1\,500\,000}{2841}-\frac{1\,500\,000\times 40}{56\,100}\right)+\frac{53\,750\,000}{56\,100}$$
$$=422-854+960 = 528 \text{ N/cm}^2$$

528 N/cm^2 < 1667 N/cm^2 which is satisfactory.

Number of strands
The characteristic breaking strength of a half-inch strand (12·5 mm) is 165 kN. It can be stressed to $0{\cdot}70\times 165 = 115{\cdot}5$ kN.

$$\left.\begin{array}{l}\text{Hence number of strands required,}\\ \text{taking a stress loss factor of}\\ 0{\cdot}80 \text{ into account}\end{array}\right\} = \frac{1500}{0{\cdot}80}\times\frac{1}{115{\cdot}5}$$

$$= 17 \text{ strands}$$

Conditions at support

(1) At transfer

$$\frac{P}{A}+\frac{Pe}{Z_b}\leqslant 1667 \qquad P = 1\,500\,000$$

$$\frac{Pe}{Z_b}\leqslant 1667-\frac{P}{A}$$

$$\leqslant 1667-\frac{1\,500\,000}{2841}$$

$$\leqslant 1667-528$$

$$\frac{1\,500\,000e}{18\,700}\leqslant 1139$$

$$e\leqslant\frac{1139\times 18\,700}{1\,500\,000}\leqslant 14{\cdot}2 \text{ cm}$$

Since the bottom kern distance is 19·75 and the eccentricity is $e = 14{\cdot}2$ cm < 19·75, there will be no tension at top fibres at transfer.

(2) Under service loads
Stress in bottom fibre will be less than that at transfer and need not be checked.
Hence eccentricity of 14 cm may be adopted.

Review of eccentricity at midspan

Because the number of strands will be about 17, their arrangement will have to be as shown in Fig. 6.8.

$$\left.\begin{array}{l}\text{Centre of gravity}\\ \text{of group from}\\ \text{bottom-most row}\end{array}\right\} = \frac{2 \times 25 + 3 \times 20 + 3 \times 15 + 3 \times 10 + 3 \times 5}{17}$$

$$= \frac{200}{17} = 11 \cdot 8$$

Available eccentricity $= 60 - 31 \cdot 8 = 28 \cdot 2$ cm
This value will have to be used to find P from the various inequalities.

(1) $\frac{1}{P} \times 10^7 \geqslant 4 \cdot 04 + 0 \cdot 94 \geqslant 4 \cdot 98$

$$P \leqslant \frac{10^4}{4 \cdot 98} \text{ kN} \leqslant 2008 \text{ kN}$$

(2) $\frac{1}{P} \times 10^7 \geqslant 0 \cdot 26e - 5 \cdot 16$

$$P \leqslant \frac{10^4}{2 \cdot 19} \text{ kN} \leqslant 4566 \cdot 21 \text{ kN}$$

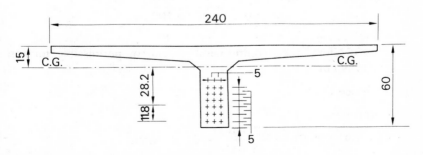

Fig. 6.8 Arrangement of tendons at midspan

(3) $\dfrac{1}{P} \times 10^7 \leqslant 0\cdot 149 e + 0\cdot 98$

$\qquad \leqslant 0\cdot 149 \times 28\cdot 2 + 0\cdot 98$

$\qquad \leqslant 4\cdot 2 + 0\cdot 98$

$\qquad \leqslant 5\cdot 18$

$\qquad P \geqslant \dfrac{10^4}{5\cdot 18} \geqslant 1930 \text{ kN}$

Number of strands required

The characteristic strength of 12·5 mm strand is 165 kN.

If it is stressed to 70%, the strength = 115·5

Number of strands required $= \dfrac{1930}{115\cdot 5} = 17$.

They will be placed as already indicated.
$\qquad$ Prestressing force $= 115\cdot 5 \times 17$

$\qquad \qquad \qquad \qquad = 1960 \text{ kN}$

Check for stresses

(1) At transfer

(a) Bottom fibre

$\qquad \dfrac{1\,960\,000}{2841} + \dfrac{1\,960\,000 \times 28\cdot 1}{18\,700} - \dfrac{32\,750\,000}{18\,700}$

$\qquad \quad 689 \quad + \quad \quad 2945 \quad - \quad \quad 1751$

$= 1883 \text{ N/cm}^2 < 2000 \text{ N/cm}^2$, which is satisfactory.

(b) At top fibre

$\qquad -\dfrac{1\,960\,000}{2841} + \dfrac{1\,960\,000 \times 28\cdot 1}{56\,100} - \dfrac{32\,750\,000}{56\,100}$

$= -692 \quad + \quad \quad 980 \quad - \quad \quad 583$

$= -295 \text{ N/cm}^2$ (i.e., compression instead of permissible tension of 100 N/cm^2)

(2) At service loads

(a) Bottom fibre

$\qquad 0\cdot 8 \times (692 + 2940) - \dfrac{53\,750\,000}{18\,700} = 31\cdot 27 \text{ N/cm}^2$

This is acceptable.

(b) Top fibre

$$0.8\left(\frac{1\,960\,000}{2841}\right)-\left(\frac{1\,960\,000\times 28\cdot 1}{56\,100}\right)$$

$$\times 0\cdot 80+\frac{53\,750\,000}{56\,100}$$

$$=552-785+958=725<1667 \text{ N/cm}^2$$

Conditions at the ends
(1) At transfer

(a) Bottom fibre

$$\frac{1\,960\,000}{2841}+\frac{1\,960\,000\times e}{18\,700}\leqslant 2000$$

$$692 \;+\; 105e \quad\;\;\leqslant 2000$$

$$105e \leqslant 1308$$

$$e \leqslant 12\cdot 45 \text{ cm}$$

(b) Top fibre

No tension will be induced as $e = 12\cdot 45$ cm will be within the kern.

Under working load, the stresses will be less and hence $e < 12\cdot 45$ cm will govern.

Let e be made $12\cdot 45$ cm at the end. Let 10 strands out of the 17 be deflected at $L/3$. Referring to Fig. 6.9, the height of the centroid of the strands from the bottom fibre is calculated as

$$\frac{2\times 50+2\times 45+2\times 40+3\times 35+3\times 30+3\times 25+1\times 10+1\times 5}{17}$$

$$= 32\cdot 65 \text{ cm}$$

$$e = 60-(32\cdot 65+15)$$

$$= 12\cdot 35 \text{ cm}.$$

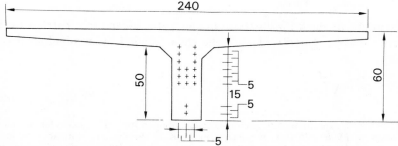

Fig. 6.9 Arrangement of tendons at end section

Check for ultimate strength

Ultimate bending moment $= 1\cdot4 \times 327\cdot5 + 1\cdot6 \times 210$ kN m
$= 458\cdot5 + 336\cdot0 = 794\cdot5$ kN m

Resisting moment of the section at midspan

$$\frac{f_{pu}A_{ps}}{f_{cu}bd} = \frac{165 \times 17\,000}{50 \times 240 \times 43\cdot2} = 0\cdot054$$

$\left(\dfrac{x}{d}\right) = 0\cdot109$ (vide Table 37 of CP 110)

$$\frac{f_{pb}}{0\cdot87 f_{pu}} = 1\cdot00$$

$M_u = 0\cdot87 f_{pu} \times A_{ps} \times (d - 0\cdot5x)$
$= 0\cdot87 \times 165 \times 17(43\cdot2 - 2\cdot34)$

$0\cdot5x = 0\cdot054 \times 43\cdot2 = 2\cdot34$

$M_u = 0\cdot87 \times 165 \times 17 \times 40\cdot86 = 99\,500$ kN cm $= 995$ kN m
995 kN m $> 794\cdot5$ kN m. Hence the section is adequate.

DESIGN FOR SHEAR*

Step 1
The first step is to assess the shear carried by nominal steel which in any case has to be provided. This assessment will be made using both the ACI and BS Code provisions.

ACI: According to ACI (26.11), the minimum web steel A_v is specified as

$$A_v = \frac{A_s}{80} \frac{f'_s}{f_y} \frac{s}{d} \sqrt{\frac{d}{b_1}}$$

According to the Mörsch truss analogy, the shear force carried by a stirrup is $\phi A_v \cdot f_y \cdot (d/s)$. Hence the shear carried by nominal steel is

$$\phi \frac{A_s}{80} f'_s \sqrt{\frac{d}{b_1}}.$$

---

* For a clear exposition of design for shear according to ACI 318:1963, reference may be made to *Modern Prestressed Concrete* by H. Kent Preston and Norman J. Sollenberger, McGraw-Hill, 1967. The author's presentation of the BS Code provisions closely follows the above presentation.

Design of pretensioned products

Applying it to the example where $\phi = 0.87$

$A_s f'_s = 17 \times 165$ kN

$d = e + y_t = 28.2 + 15 = 43.2, \quad b_1 = 20$

Hence the shear carried by nominal steel $= \dfrac{17 \times 165}{80} \sqrt{\dfrac{43.2}{20}} \times 0.87$

$= 44.4$ kN

BS Code: According to the BS Code the spacing of stirrups shall not exceed 0·75 times the effective depth of the beam. Hence provide stirrups at 0.75×43.2, i.e., at a spacing of 32·4 cm or say 32 cm.

Area of mild-steel stirrups $= 0.002 b_t \times s_v$
$= 0.002 \times 20 \times 32$
$= 1.28$ cm^2

$\left.\begin{array}{l}\text{Shear force carried by}\\ \text{nominal reinforcement}\end{array}\right\} = 1.28 \times f_y \times 0.87 \times \dfrac{43.2}{32.0}$

$f_y = 250$ N/mm^2

$\left.\begin{array}{l}\text{Shear carried by}\\ \text{nominal reinforcement}\end{array}\right\} = 1.28 \times 25 \times 0.87 \times \dfrac{43.2}{32}$

$= 37.60$ kN

There is fair agreement between the two.

Step 2: Computation of ultimate shear V_u

$V_u = 1.4$ (D.L. shear) $+ 1.6$ (L.L. shear)

D.L. shear $= 65.5$ kN

L.L. shear $= 42$ kN

V_u at support $= (1.4 \times 65.5 + 1.6 \times 42) = (92 + 67.2)$
$= 159.2$ kN

$\left.\begin{array}{l}\text{Maximum shear permissible}\\ \text{at support}\end{array}\right\} = 27.35 \times 20 \times 530 = 290\,000$ N

$= 290$ kN

$159.2 < 290$, which is satisfactory.

Step 3: Computation of v_{co}

$V_{co} = 0.67 bh \sqrt{f_t^2 + 0.8 f_{cp} f_t}$

f_{cp} = compressive stress at centroidal axis

$f_{cp} = \dfrac{0.80 \times 1960}{2841} = 0.552 \text{ kN/cm}^2$

$\qquad = 552 \text{ N/cm}^2$

$\qquad = 5.52 \text{ N/mm}^2$

	V_{co}/bh
For 50 grade concrete with $f_{cp} = 6$ (Ref: Table 39, CP 110:1972)	2.20
For 50 grade concrete with $f_{cp} = 4$	1.95
Difference	0.25

Hence, proportionately for a difference of 1.5, f_{cp} $\bigg\} = \dfrac{0.25}{2} \times \dfrac{3}{2} = 0.19$

Hence $\dfrac{V_{co}}{bh}$ for $f_{cp} = 5.5$ is $= 1.95 + 0.19 = 2.14$

Hence $V_{co} = bh \times 2.14 = 20 \times 60 \times 2.14 \times \dfrac{100}{1000}$

$\qquad = 256.8 \text{ kN}$

The vertical component of 15 inclined cables has to be added. Referring to Fig. 6.10,

vertical component $= 1960 \times \dfrac{15}{17} \times \tan \theta$

$\qquad = 1960 \times \dfrac{15}{17} \times \dfrac{15 \cdot 85}{667} = 41 \text{ kN}$

Step 4: Computation of V_{cr}

(a) V_{cr} (Sections cracked in flexure)

$V_{cr} = \left(1 - 0.55 \dfrac{f_{pe}}{f_{pu}}\right) v_c bd + M_0 \dfrac{V}{M}$

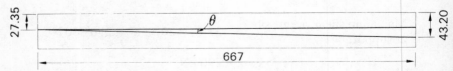

Fig. 6.10 Details of draped tendons

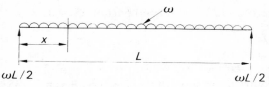

Fig. 6.11 Shear and bending moment at a section

(b) V/M for uniformly distributed loads

Shear, V at x (Fig. 6.11) $= \dfrac{w}{2}(L-2x)$

B.M., M at x $\quad = \dfrac{w}{2}(Lx-x^2)$

$$\dfrac{V}{M} = \dfrac{L-2x}{Lx-x^2}$$

Values of V/M at different sections are tabulated in Table 6.2.
M_0 = moment at which there is zero stress in concrete at a depth d.

(1) At $L/2$,

$M_0 = 0{\cdot}8 \times f_{pt}(I/y)$

$f_{pt} = 0{\cdot}80 \times 1960\left(\dfrac{1}{2841} + \dfrac{28{\cdot}2}{18\,700}\right) \times 1000$

$\quad = 2916 \text{ N/cm}^2$

$M_0 = \dfrac{0{\cdot}80 \times 2916}{10^5} \times \dfrac{18\,700 \times 45}{43{\cdot}2}$ kN m

$\quad = 455$ kN m

Table 6.2

x	$\dfrac{V}{M}$
$\dfrac{L}{8}$	$\dfrac{6{\cdot}85}{L}$
$\dfrac{L}{4}$	$\dfrac{2{\cdot}67}{L}$
$\dfrac{L}{3}$	$\dfrac{1{\cdot}50}{L}$
$\dfrac{L}{2}$	0

76 Modern prestressed concrete design

V/M being zero at $L/2$, the second part of $V_{cr} = 0$.

(2) At $L/8$
Eccentricity at $L/8$
Eccentricity at $(L/3 = 28\cdot 2) = e + y_t = 43\cdot 2$
Eccentricity at support $= (12\cdot 35) = e + y_t = 27\cdot 35$
Difference $= 15\cdot 85$

Hence, $e + y_t$ at $L/8$ (by interpolation)

$$= 27\cdot 35 + \left(\frac{250}{1000} \times \frac{3}{2}\right) \times 15\cdot 85$$

$$= 27\cdot 35 + 5\cdot 96 = 33\cdot 31 = e + y_t = d$$

$$e = 33\cdot 31 - 15 = 18\cdot 31$$

$$M_0 = 0\cdot 80 f_{pt}(I/y)$$

$$f_{pt} = 0\cdot 80 \times 1960 \left(\frac{1}{2841} + \frac{18\cdot 31}{18\,700}\right) \times 1000$$

$$= 550 + 1535 = 2085$$

$$M_0 = \frac{0\cdot 80 \times 2085}{10^5} \times \frac{18\,700 \times 45}{33\cdot 3} \text{ kN m}$$

$$= 420 \text{ kN m}$$

$$\frac{M_0 V}{M} = \frac{420 \times 6\cdot 85}{20} = 144 \text{ kN}$$

(3) At $L/4$
Eccentricity at $L/4$

$$e + y_t = 27\cdot 35 + \frac{3}{4} \times 15\cdot 85 = 39\cdot 25 = d$$

$$e = 24\cdot 25$$

$$f_{pt} = 0\cdot 80 \times 1960 \left(\frac{1}{2841} + \frac{24\cdot 25}{18\,700}\right) \times 1000$$

$$= (550 + 2040) = 2590 \text{ N/cm}^2$$

$$M_0 = \frac{0\cdot 80 \times 2590}{10^5} \times \frac{18\,700 \times 45}{39\cdot 25}$$

$$= 443 \text{ kN m}$$

$$M_0 \frac{V}{M} = 443 \times \frac{8}{60} = \frac{354\cdot 4}{6} = 59\cdot 0 \text{ kN}$$

(4) At $L/3$

$M_0 = 455$ kN m

$M_0 \dfrac{V}{M} = \dfrac{455 \times 3}{2 \times 20} = 34$ kN

(c) Evaluation of term

$$\left(1 - 0{\cdot}55 \dfrac{f_{pe}}{f_{pu}}\right) v_c bd$$

$\dfrac{f_{pe}}{f_{pu}} = \dfrac{115{\cdot}5 \times 0{\cdot}80}{165} = 0{\cdot}56 < 0{\cdot}60$, which is satisfactory.

$1 - 0{\cdot}56 \times 0{\cdot}55 = 0{\cdot}692, \quad v_c = 0{\cdot}92$ N/mm^2.

Table 6.3

Section	$0{\cdot}44\, v_c bd$ in kN
$\dfrac{L}{8}$	42·5
$\dfrac{L}{4}$	50·0
$\dfrac{L}{3}$	55·0
$\dfrac{L}{2}$	55·0

Note
 For finding v_c, $100 A_s/bd$ will be taken to correspond to the central section as this will be conservative.

$\dfrac{100 \times 17 \times 94{\cdot}2}{20 \times 43{\cdot}2 \times 100} = 1{\cdot}85$

For $\dfrac{A_s \times 100}{bd}$ of 1·00   0·75

For $\dfrac{A_s \times 100}{bd}$ of 2·00   0·95

Difference   0·20

For a difference of $0{\cdot}85 = 0{\cdot}20 \times 0{\cdot}85 + 0{\cdot}75$
$\phantom{For a difference of 0{\cdot}85} = 0{\cdot}92$ N/mm^2

Minimum $V_{cr} = 0{\cdot}10 \times bd \times \sqrt{f_{cu}}$

$\phantom{Minimum V_{cr}} = 0{\cdot}10 \times 20 \times 43{\cdot}2 \times 100 \sqrt{50}$ N

$\phantom{Minimum V_{cr}} = 61$ kN

78 Modern prestressed concrete design

Step 5: Conditions at support
Transfer length of half-inch strand = 330 mm
Let member overhang 7·5 cm beyond centre of support (Fig. 6.12).

$$f_{cp} \text{ at support} = \frac{7\cdot 5}{33} \times 5\cdot 52 = 1\cdot 26 \text{ N/mm}^2$$

At support

$$V_{co} = 0\cdot 67 bh\sqrt{f_t^2 + 0\cdot 8 f_{cp} \cdot f_t}$$

Referring to Table 39, CP 110:1972

$$\text{for } f_{cp} = 2, \quad \text{value of } \frac{V_{co}}{bh} = 1\cdot 60$$

It is, therefore, better to calculate and find the value

$$V_{co} = 100 \times 0\cdot 67 \times 20 \times 60\sqrt{0\cdot 24 \times 50 + 0\cdot 8 \times 1\cdot 26 \times 0\cdot 24 \times 50}$$

$$= 1200 \times 100 \times \tfrac{2}{3}\sqrt{1\cdot 68^2 + 0\cdot 80 \times 1\cdot 26 \times 1\cdot 68}$$

$$= 80\,000\sqrt{2\cdot 85 + 1\cdot 7} \times 10^{-3} \text{ kN}$$

$$= 80\sqrt{4\cdot 55} = 80 \times 2\cdot 14 = 171\cdot 2 \text{ kN}$$

This may be checked by using Table 39, CP 110:1972

$$\text{For } f_{cp} = 0, \quad \frac{V_{co}}{bh} = 0\cdot 67 \times f_t$$

$$= 0\cdot 67 \times 0\cdot 24\sqrt{50}$$

$$= 1\cdot 1280$$

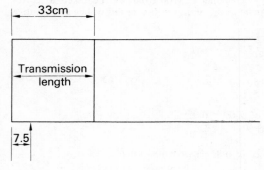

Fig. 6.12 Transmission length

Design of pretensioned products

$$f_{cp} = 0, \quad \frac{V_{co}}{bh} = 1\cdot 13 \quad f_{cp} = 2, \quad \frac{V_{co}}{bh} = 1\cdot 60$$

For $f_{cp} = 1\cdot 26$, $\quad \dfrac{V_{co}}{bh} = 1\cdot 13 + \dfrac{0\cdot 47}{2} \times \dfrac{5}{4} = 1\cdot 43$

$V_{co} = 143 \times 20 \times 60 \times 10^{-3} = 172$ kN

Step 6

From Table 6.4, it is evident that V_{co} or V_{cr} is higher than $V_u - V'_u$ at every section. Hence no shear reinforcement is necessary. However, nominal two-legged stirrups of 10 mm diameter with an area of $1\cdot 57$ cm² shall be provided at a spacing not exceeding $0\cdot 75$ times the effective depth which is $0\cdot 75 \times 43\cdot 2 = 32\cdot 4$ cm, say 32 cm. The code requires a minimum area of stirrups $= \dfrac{0\cdot 40 \times b \times s_v}{0\cdot 87 f_y V}$

$= \dfrac{0\cdot 40 \times 20 \times 32\cdot 4}{0\cdot 87 \times 250 \times 100} = 1\cdot 20$ cm²

Hence the stirrups provided are adequate.

Instantaneous Camber

Prestressing force before losses = 1960 kN
Pe at support = $1960 \times 12\cdot 35$ kN cm = 24 200 kN cm
Pe at $L/3$ = $1960 \times 28\cdot 2 = 55\,100$ kN cm

Table 6.4 Shearing force and shearing resistance (in kN)

Station	Distance from support cm	V_u	V'_u	$V_u - V'_u$	V_{co}	V_{cr}	Remarks
	0	159·2	37·60	121·60	172·00		
	25·5	155·5	37·60	117·90	298·00		
L/8	250	119·4	37·60	81·80	298·00	182·5	
L/4	500	79·6	37·60	42·00	298·00	109·0	
L/3	667	53·00	37·60	15·40	298/257	89·0	
L/2	1000	0·00	37·60		257	61·0 kN*	*Minimum of $0\cdot 10\, bd\, \sqrt{f_{cu}}$ governs

Notes:
V'_u = Shear carried by nominal stirrups
Transmission length = 33 cm. It is assumed that the prestress varies from 0 to its full value linearly from its free edge to the point 33 cm away from it. Beam overhangs centre of supports by 7·5 cm.

Actually, 24 200 will be effective only at a distance of 33 cm from the end which is the transmission length. But the error involved in assuming that a moment of 24 200 is effective at the support is negligible. Deflexion at midspan may be computed by using the area-moment theorem and the bending-moment diagram given in Fig. 6.13.

Moment of bending moment deflexion about support

$= (55\,100 \times 333 \times 834) + (0{\cdot}5 \times 667 \times 30\,900 \times 444)$
$\quad + (24\,200 \times 667 \times 333)$
$= (1530 \times 10^7 + 460 \times 10^7 + 538 \times 10^7)$
$= 2528 \times 10^7$ kN cm

$I_{xx} = 842\,474$ cm^4

$E_c = 34$ kN/mm^2 for instantaneous deflexion

Camber (initial) due to prestress $= \dfrac{2528 \times 10^7}{842\,474 \times 3400}$

$= \dfrac{2528}{84{\cdot}25 \times 3{\cdot}4} = 8{\cdot}83$ cm

Dead-load deflexion $= \dfrac{5}{384} \times \dfrac{6{\cdot}55 \times 16 \times 10^4 \times 10^6}{3400 \times 842\,474}$

$= 4{\cdot}76$ cm

Net upward camber $= 4{\cdot}07$ cm

Camber to be limited to $\dfrac{1}{300} \times$ span

$\dfrac{\text{Span}}{300} = \dfrac{2000}{300} = 6{\cdot}67$ cm, which is satisfactory.

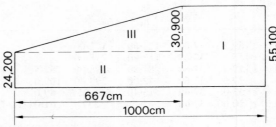

Fig. 6.13 Bending moment diagram

Design Example 6.2
Pretensioned hyperboloid roofing elements of 18 m span and 18·75 m total length are to be standardized for manufacture. The characteristic

Design of pretensioned products 81

cube strength of concrete specified is 50 N/mm² and the transfer cube strength is 40 N/mm². The tendons are to consist of 5 mm wires. The prestressing losses are estimated at 20%. The available crane capacity is 7 t. $\gamma = 2500$ kg/m³.

Step 1

$$w = \frac{1000 \times 7}{2 \cdot 2956 \times 18 \cdot 75 \times 2500 \times 0 \cdot 06} = 1 \cdot 08 \text{ m}$$

Adopt $w = 1 \cdot 05$ m

Width of unit $2w = 2 \cdot 10$ m. The section may now be sketched out (Fig. 6.14).

Properties of unit using (6.6) to (6.11):

Area $A_c = 1446$ cm²

$Y_b = 19 \cdot 22$ cm

$Z_b = 19\,428$ cm³

$Z_t = 11\,219$ cm³

Loads and moments

Self-weight of unit	$= 3 \cdot 62$ kN/m
Live load on unit	$= 2 \cdot 10$ kN/m
Total load	$= 5 \cdot 72$ kN/m

$$\text{Total bending moment} = \frac{5 \cdot 72}{2} \times 18 \cdot 75 \left(9 - \frac{9 \cdot 375}{2}\right)$$

$$= 231 \cdot 26 \text{ kN/m}$$

$$\text{Dead load moment} = \frac{3 \cdot 62 \times 18 \cdot 75}{2} \left(9 - \frac{9 \cdot 375}{2}\right)$$

$$= 146 \cdot 36 \text{ kN m}$$

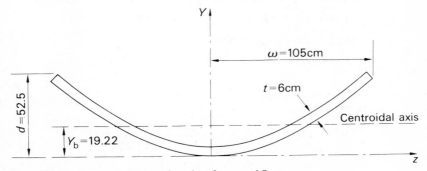

Fig. 6.14 Cross-section of unit of span 18 m

Step 2

$$e = 19\cdot 22 - 4\cdot 15 = 15\cdot 07 \text{ cm}$$

Step 3

Effective prestressing force P_e is found from (6.12) as

$$P_e = \frac{231\cdot 260 \times 100}{\left(\dfrac{19\,428}{1446} + 15\cdot 07\right)} = 811 \text{ kN}$$

$$P = 811 \times 1\cdot 25 \quad = 1014 \text{ kN}$$

$$\left.\begin{array}{l}\text{70\% characteristic}\\\text{strength of stress-}\\\text{relieved 5 mm wires}\end{array}\right\} = 0\cdot 70 \times 30\cdot 8 = 21\cdot 56 \text{ kN}$$

Hence, number of 5 mm wires required $= \dfrac{1014}{21\cdot 56} = 47\cdot 03$

Use 48 wires

Step 4

Check for stresses at midspan

After losses, 48 wires of 5 mm will provide an effective prestress of 827 727 kN.

(1) Stress at bottom, under service loads

$$= \frac{827\,727}{1446} + \frac{827\,727 \times 15\cdot 07}{19\,428} - \frac{231\,260 \times 100}{19\,428}$$

$$= 24 \text{ N/cm}^2 = 0\cdot 24 \text{ N/mm}^2 \text{ (compression)}$$

(2) Stress at top, under service loads

$$= \frac{827\,727}{1446} + \frac{231\,260 \times 100}{11\,219} - \frac{827\,727 \times 15\cdot 07}{11\,219}$$

$$= 1522 \text{ N/cm}^2 = 15\cdot 22 \text{ N/mm}^2$$

At transfer

(3) Stresses at midspan
    At top fibre

$$= \frac{1\,034\,658}{1446} - \frac{1\,034\,658 \times 15\cdot 07}{11\,219} + \frac{146\,360 \times 100}{11\,219}$$

$$= 630 \text{ N/cm}^2 = 6\cdot 30 \text{ N/mm}^2 \text{ (compression)}$$

Design of pretensioned products

At bottom fibre

$$= \frac{1\,034\,658}{1446} + \frac{1\,034\,658 \times 15\cdot07}{19\,428} - \frac{146\,360 \times 100}{19\,428}$$

$= 765 \text{ N/cm}^2 = 7\cdot65 \text{ N/mm}^2$ (compression)

Step 5
Check for stresses at end section

Using equation (6.15), $a = \dfrac{L^2}{8e} = \left(\dfrac{18\cdot75 \times 18\cdot75}{8 \times 0\cdot1507}\right)$

$\qquad\qquad\qquad\qquad\qquad = 291\cdot6$ m

Let us use a radius $a = 225$ m

End eccentricity $= 15\cdot07 - \dfrac{18\cdot75 \times 18\cdot75}{8 \times 225} \times 100$

$\qquad\qquad\qquad = 15\cdot07 - 19\cdot53$

$\qquad\qquad\qquad = -4\cdot46$ cm (i.e., above c.g. of section)

(1) Stress at top at transfer $= \dfrac{1\,034\,658}{1446} + \dfrac{1\,034\,658 \times 4\cdot46}{11\,219}$

$\qquad\qquad\qquad = 1127 \text{ N/cm}^2$

$\qquad\qquad\qquad = 11\cdot27 \text{ N/mm}^2$

(2) Stress at bottom fibre $= \dfrac{1\,034\,658}{1446} - \dfrac{1\,034\,658 \times 4\cdot46}{19\,428}$

$\qquad\qquad\qquad = 478 \text{ N/cm}^2$

$\qquad\qquad\qquad = 4\cdot78 \text{ N/mm}^2$

All stresses are within permissible limits.

Step 6
Check for ultimate moment of resistance

Ultimate design moment $= (1\cdot4 \times 146\cdot36 + 1\cdot6 \times 84\cdot9)$

$\qquad\qquad\qquad\qquad = 340\cdot74$ kN m

Assume a trial neutral axis depth of 32 cm. The area in compression $= 2 \times 52\cdot75 \times 6 = 633$ cm^2.

Total compression $= 633 \times 0\cdot40 \times f_{cu}$

$\qquad\qquad\qquad = 633 \times 0\cdot40 \times \dfrac{5000}{1000} = 1266$ kN

84 Modern prestressed concrete design

It is easily verified that the steel reaches its ultimate strength. Hence,

$$\text{total tension} = \frac{30\cdot 8 \times 48}{1\cdot 15} = 1285 \text{ kN}$$

The total tension and compression are nearly equal. Hence, the assumed position of the neutral axis is nearly correct. Assuming the centre of compression to be located at $0\cdot 40 \times 32 = 12\cdot 8$ cm from the top, the lever arm $= 48\cdot 35 - 12\cdot 8 = 35\cdot 55$ cm. Hence, the resisting moment

$$= 1266 \times \frac{35\cdot 55}{100} = 450 \text{ kN m} > 340\cdot 74 \text{ kN m}.$$

Hence, the structure is safe.

Step 7
Check for shear

$$\left.\begin{array}{l}\text{Ultimate shear due to}\\ \text{self-weight}\end{array}\right\} = \frac{3\cdot 62 \times 18\cdot 75}{2} \times 1\cdot 4 = 47\cdot 5 \text{ kN}$$

$$\left.\begin{array}{l}\text{Ultimate shear due to}\\ \text{live load}\end{array}\right\} = \frac{2\cdot 10 \times 18\cdot 75}{2} \times 1\cdot 6 = 31\cdot 5 \text{ kN}$$

Total ultimate shear $= 79$ kN

Maximum shear stress $= \dfrac{79\,000}{144\,600} = 0\cdot 55 \text{ N/mm}^2$

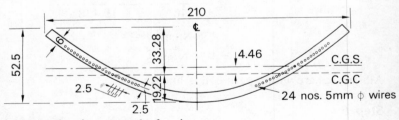

Fig. 6.15(a) Section at end of unit

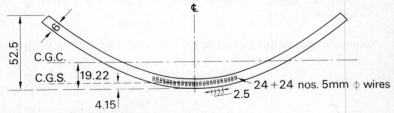

Fig. 6.15(b) Section at midspan

Design of pretensioned products 85

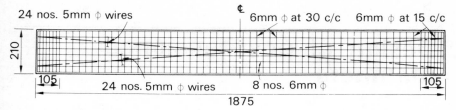

Fig. 6.15(c) Plan of unit

$\left.\begin{array}{l}\text{Allowable shear stress}\\ v_c \text{ for } f_{cu} = 50 \text{ N/mm}^2 \\ \text{and with } 0\cdot65\% \text{ steel}\end{array}\right\} = 0\cdot60 \text{ N/mm}^2 \text{ (vide Table 5, CP 110:1972)}$

Hence, a nominal reinforcement consisting of eight 12 mm bars in the longitudinal direction and 6 mm bars at 30 cm centre to centre in the transverse direction will do. The details of prestressing tendons and nominal reinforcement provided may be seen in Fig. 6.15.

References

[6.1] PCI Design Handbook: *Precast and Prestressed Concrete*, Prestressed Concrete Institute, Chicago, Illinois.

[6.2] Rowe, R. E., *Concrete Bridge Design*, Chapter 11, Wiley, New York, 1962.

[6.3] *HP-Schalen Heft*, NORMCO, Gesellschaft für Normkonstruktionen und Statik mbH, Essen.

[6.4] Rahman, Abdul, P. M., Deshmukh, R. S., and Jain, S. S., Precast prestressed hyperboloid shells for industrial roofs, *Indian Concrete Journal*, November, 1968.

[6.5] Ramaswamy, G. S., and Sayeed, N. R., 'Simplified analysis of pretensioned hyperboloid roofing units', bulletin of the International Association for Shell and Spatial Structures, n. 55, Madrid (Spain).

7
Design of post-tensioned members

Post-tensioning has distinct advantages over pretensioning for members of long span carrying heavy loads. Such members tend to be too long or too heavy to be produced in the factory and transported to the site. Post-tensioning permits the member to be cast *in situ* and stressed. The ducts to house the tendons are preformed during casting by embedding corrugated sheaths. The member is post-tensioned after the concrete has sufficiently hardened. Subsequently, the tendons are protected by grouting the ducts.

7.1 Segmental Construction

It is sometimes advantageous to cast a girder in segments and post-tension it, the joints between the segments being formed by concrete or epoxy mortar [7.1] and [7.2]. Such construction is especially advantageous in out of the way locations with insufficient facilities for *in situ* work. This technique also finds application in the free-cantilever method of post-tensioned bridge construction [7.3]. The author has successfully used the following epoxy mortar formulation suggested by CIBA-GEIGY of India.

$$\text{Binder} \begin{cases} \text{Resin (Araldite GY257), 24\% by weight} \\ \text{Hardener (X157/2401), 4\% by weight} \end{cases}$$

$$\left.\begin{array}{l}\text{Filler—A mixture of quartz sand}\\ \text{and silica in the ratio}\\ \text{of 3:7}\end{array}\right\} 72\% \text{ by weight}$$

The pot-life of the mixture was 20 to 25 minutes at an ambient temperature of 30°C. The epoxy mortar is usually applied to a thickness of about 1·5 mm on the abutting faces of adjacent segments. More detailed information is available in [7.1].

7.2 Design Procedure

The design procedure for post-tensioned beams is explained in detail in the worked example that follows. The example deals with the design of a composite post-tensioned bridge-girder (unpropped).

Design Example 7.1

Data

Span = 30 m centre to centre of bearings
Live loads: (a) Uniformly distributed live load of 18·85 kN/m
(b) A knife-edge load of 78·26 kN
System of prestressing: CCL-Gifford Udall System
Thickness of top slab = 15 cm
Characteristic strength of concrete in girder = 40 N/mm^2
Characteristic strength of concrete in slab = 25 N/mm^2

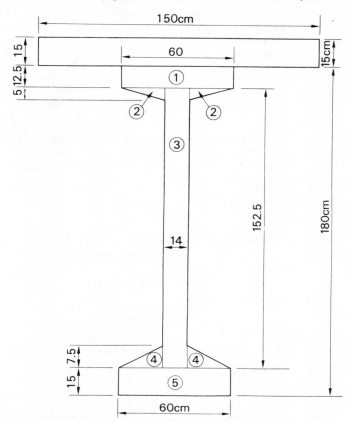

Fig. 7.1 Cross-section of composite bridge girder

88 Modern prestressed concrete design

(1) Method

A trial section is assumed as shown in Fig. 7.1. The properties of the girder and the composite section are next computed.

(2) Properties of section

Area = 4072·5 cm^2

y_t = 93·0827 62 cm

y_b = 86·917 24 cm

Moment of inertia of girder about top (See Table 7.1) $\Big\}$ = 52 369 563 cm^4

Table 7.1 Computation of I of girder about top

No	Area	Moment of inertia about own axis	d	d^2	Ad^2	Total moment of inertia
1	750	9766	6·25	39·06	29 295	39 061
2	57·5	80	14·17	200·79	11 545	11 625
2	57·5	80	14·17	200·79	11 545	11 625
3	2135	41 37 674	88·75	7 876·56	16 816 460	20 954 134
4	86·25	270	162·5	26 406·25	22 77 539	2 277 809
4	86·25	270	162·5	26 406·25	22 77 539	2 277 809
5	900	16 875	172·5	29 756·25	26 780 625	26 797 500
	4072·5	41 65 015			48 204 548	52 369 563

Moment of inertia of girder about its c.g. $\Big\{$ $\begin{array}{l} 52\ 369\ 563 \\ -4072 \cdot 5\ (93 \cdot 0827\ 62)^2 \end{array}$

= 17 083 792 cm^4

Z_t = $\dfrac{17\ 083\ 792}{93 \cdot 0827\ 62}$

= 183 533·35 cm^3

Z_b = $\dfrac{17\ 083\ 792}{86 \cdot 917\ 24}$

= 196 552·39 cm^3

(3) *Properties of composite section*

Moment of inertia of slab about own axis $= \frac{1}{12} \times 150 \times 15^3$

$= 42\,188 \text{ cm}^4$

Transformed moment of inertia of slab $= 0.80 \times 42\,188$

$= 33\,750 \text{ cm}^4$

$y_t^c =$ distance to top of composite section

$= \dfrac{(1800 \times 7.5) + (4072.5 \times 108.082\,762)}{5872.5}$

$= 77.252\,795$

$y_b^c =$ distance to bottom fibre of composite section

$= 195 - 77.252\,795 = 117.747\,21$

Moment of inertia of composite section

$= 17\,083\,972 + 4072.5(93.082\,762 - 77.252\,795)^2$

$\quad + 33\,750 + 1800(77.252\,795 - 7.5)^2$

$= 26\,895\,875 \text{ cm}^4$

$Z_t^c = \dfrac{26\,895\,875}{77.252\,795} = 348\,154.05 \text{ cm}^3$

$Z_b^c = \dfrac{26\,895\,875}{117.7472} = 228\,420.51 \text{ cm}^3$

(4) *Bending moments*

Weight of girder $= 0.407\,25 \times 24 = 9.77$ kN/m

B.M. due to girder $= \dfrac{9.77 \times 30 \times 30}{8} = 1099.12$ kN m

Weight of slab $= 0.225 \times 24 = 5.4$ kN/m

B.M. due to slab $= \dfrac{5.4 \times 30 \times 30}{8} = 607.5$ kN m

B.M. due to girder and slab $= 1706.625$ kN m

B.M. due to uniformly distributed live load $= \dfrac{18.85 \times 30 \times 30}{8}$

$= 2120.625$ kN m

90 Modern prestressed concrete design

B.M. due to knife-edge load $= \dfrac{78 \cdot 26 \times 30}{4}$

$\qquad\qquad\qquad\qquad\qquad\qquad = 586 \cdot 95$ kN m

Total live load B.M. $\qquad\quad = 2707 \cdot 575$ kN m

B.M. due to asphalt and railing $= \dfrac{3 \times 30 \times 30}{8}$

$\qquad\qquad\qquad\qquad\qquad\qquad = 337 \cdot 5$ kN m

B.M. on composite section $\quad = 3045 \cdot 09$ kN m

(5) Stresses

Stress in bottom fibre due to girder and slab $= \dfrac{1706 \cdot 625 \times 1000}{196\,552}$

$\qquad\qquad\qquad\qquad\qquad\qquad\quad = 8 \cdot 60$ N/mm^2

Stress in bottom fibre of composite girder due to live load, railing and asphalt $= \dfrac{3\,045\,090}{228\,420 \cdot 51} = 13 \cdot 33$ N/mm^2

Total stress in the bottom fibre $= 13 \cdot 33 + 8 \cdot 60 = 21 \cdot 93$ N/mm^2

(6) Preliminary estimate of prestressing force

Let us provide 42 tendons, in six groups of 7 strands of 12·5 mm each. It is assumed that the girder is post-tensioned at 28 days. The available $e = 66 \cdot 92$ cm after allowing for adequate concrete cover and location of strands.

Check for stresses

Stress in bottom fibre due to girder moment $= \dfrac{1\,099\,120}{196\,552} = 5 \cdot 59$ N/mm^2 (tension)

Stress in bottom fibre due to prestress, each cable being prestressed to $0 \cdot 70 \times 165 = 115 \cdot 5$ kN $= \dfrac{42 \times 115 \cdot 5 \times 1000}{100} \left(\dfrac{1}{4072 \cdot 5} + \dfrac{66 \cdot 92}{196\,552} \right)$

$= 28 \cdot 43$ N/mm^2 (compression)

Net stress in bottom fibre at transfer $= 28 \cdot 43 - 5 \cdot 59 = 22 \cdot 83$ N/mm^2

against 20 N/mm^2 permissible. This may be regarded as admissible as the fibre will be relieved as soon as the top slab is cast.

Design of post-tensioned members

Stress in top fibre due to girder moment $= \dfrac{1\,099\,125}{183\,533}$

$= 5 \cdot 99 \text{ N/mm}^2 \text{ (compression)}$

Stress in top fibre due to prestress $= \dfrac{115 \cdot 5 \times 42 \times 1000}{100}$

$\times \left(\dfrac{1}{4072 \cdot 5} - \dfrac{66 \cdot 92}{183\,533} \right)$

$= -5 \cdot 78 \text{ N/mm}^2 \text{ (tension)}$

net stress in top fibre $= 5 \cdot 99 - 5 \cdot 78$

$= 0 \cdot 21 \text{ N/mm}^2 \text{ (compression)}$

Stress in bottom fibre of girder due to slab $= \dfrac{607\,500}{196\,552} = 3 \cdot 09 \text{ N/mm}^2$

Hence net stress in bottom fibre after slab is cast $= 22 \cdot 83 \times 0 \cdot 80 - 3 \cdot 09$

$= 15 \cdot 17 \text{ N/mm}^2 \text{ (compression)}$

Stress in top fibre due to slab $= \dfrac{607\,550}{183\,533}$

$= 3 \cdot 31 \text{ N/mm}^2 \text{ (compression)}$

Net stress in top fibre $= 3 \cdot 31 + 0 \cdot 8 \times 0 \cdot 21$

$= 3 \cdot 48 \text{ N/mm}^2 \text{ (compression)}$

Stress in top fibre of composite slab due to live load, asphalt and railing $= \dfrac{3\,045\,090}{348\,154 \cdot 05}$

$= 8 \cdot 75 \text{ N/mm}^2 \text{ (compression)}$

Stress at bottom fibre of slab $= \dfrac{3\,045\,090}{26\,862\,125} \times 62 \cdot 252\,795$

$= 7 \cdot 06 \text{ N/mm}^2 \text{ (compression)}$

Stress at top of girder due to all loads $= 7 \cdot 06 + 3 \cdot 48$

$= 10 \cdot 54 \text{ N/mm}^2 \text{ (compression)}$

These stresses are within permissible limits.

92 Modern prestressed concrete design

$$\left.\begin{array}{l}\text{Stress in bottom fibre}\\ \text{of composite section due}\\ \text{to liveload, asphalt and}\\ \text{railing}\end{array}\right\} = \frac{3\,045\,090}{228\,420 \cdot 51}$$

$$= 13 \cdot 33 \text{ N/mm}^2 \text{ (tension)}$$

$$\left.\begin{array}{l}\text{Net stress in bottom}\\ \text{fibre of composite section}\end{array}\right\} = 15 \cdot 17 - 13 \cdot 33$$

$$= 1 \cdot 84 \text{ N/mm}^2 \text{ (compression)}$$

Hence the prestressing force and eccentricity assumed are suitable.

(7) *Arrangement of tendons at the central section and at the ends*

The tendons will be arranged at the central section as indicated in Fig. 7.2. A rectangular end-block will be provided and the strands will be anchored at the ends as shown in Fig. 7.3.

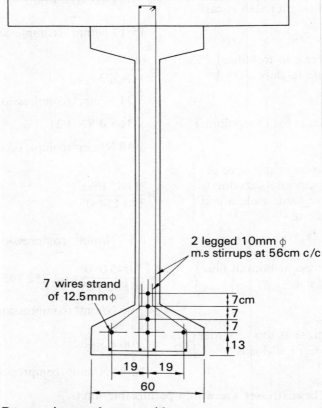

Fig. 7.2 Prestressing tendons at midspan

Design of post-tensioned members

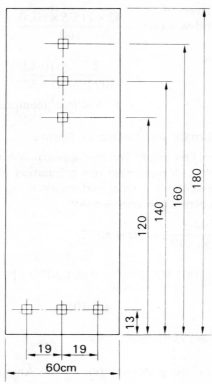

Fig. 7.3 Anchorages at end section

The height of the c.g. of the tendons above the bottom fibre at the end section
$$= \frac{3 \times 13 + 120 + 140 + 160}{6}$$
$$= 76.5 \text{ cm}$$

Eccentricity of the prestressing force at the end section $= 86.92 - 76.50 = 10.42$ cm

Compressive stress at the bottom fibre of end section at transfer due to prestress
$$= \frac{42 \times 115.5 \times 1000}{100}$$
$$\times \left(\frac{1}{4072.5} + \frac{10.42}{196\,552}\right)$$
$$= 14.48 \text{ N/mm}^2 < 20 \text{ N/mm}^2$$

Hence the arrangement is acceptable.

Compressive stress at top fibre of end-section at transfer $= \dfrac{42 \times 115 \cdot 5 \times 1000}{100}$

$$\times \left(\dfrac{1}{4072 \cdot 5} - \dfrac{10 \cdot 42}{183\,533} \right)$$

$= 9 \cdot 16 \text{ N/mm}^2 \text{ (compression)}$

(8) Check for ultimate limit state of flexure

The check will be first made by the approximate method given in ACI Code 318:1962. A more rigorous calculation will be made later by computing stresses in the steel and concrete and considering the strain distribution across the cross-section.

$$\dfrac{A_{ps}}{bd} = \dfrac{42 \times 94 \cdot 2}{100 \times 150 \times 175} = 0 \cdot 001\,507\,2$$

$$\text{Steel index} = 0 \cdot 001\,507\,2 \times \dfrac{f_y}{f_{cu}} = 0 \cdot 001\,507\,2 \times \left(\dfrac{1751}{25} \right)$$

$$= 0 \cdot 1056$$

$$f_y = \dfrac{165\,000}{94 \cdot 2} = 1751 \text{ N/mm}^2$$

$f_{cu} = 25 \text{ N/mm}^2$ using the notation of the American Code,

$$A_{sf} = \dfrac{0 \cdot 45 \times 25 \times (100 - 14) \times 15}{1751} = 13 \cdot 107 \text{ cm}^2$$

= area of steel for developing the ultimate strength of the overhanging flange

$$A_{sr} = A_s - A_{sf} = \dfrac{42 \times 94 \cdot 2}{100} - 13 \cdot 07 = 26 \cdot 45 \text{ cm}^2.$$

Ultimate moment of resistance contributed by the overhanging flange

$$= \dfrac{0 \cdot 45 \times 2500 \times 136(175 - 7 \cdot 5)}{1000 \times 100} = 3844 \text{ kN m.}$$

Ultimate moment of resistance contributed by the steel is

$$\dfrac{26 \cdot 45}{100} \times \dfrac{175 \cdot 1}{1 \cdot 15} \times 175 \left(1 - \dfrac{0 \cdot 59 \times 26 \cdot 45 \times 1751}{14 \times 175 \times 25} \right) = 3900 \text{ kN m.}$$

Design of post-tensioned members

This will have to be multiplied by the steel stress factor which is $(1-0\cdot 50\times 0\cdot 1056)=0\cdot 9472$. Hence the steel contribution is 3694·08 kN m.

Total ultimate moment of resistance $= 3844+3694\cdot 08$

$$= 7538\cdot 08 \text{ kN m.}$$

Ultimate bending moment $= 1\cdot 4\times 2044\cdot 125+1\cdot 6\times 2707\cdot 575$

$$= 7193\cdot 895 \text{ kN m.}$$

Hence, the ultimate moment of resistance is greater than the ultimate bending moment.

Alternative method

Let the neutral axis be assumed to lie 95 cm below the top fibre.

Strain at the level of steel corresponding to maximum compressive strain in concrete at top $\Bigg\} = 0\cdot 0035 \times \dfrac{100}{95} = 0\cdot 003\,684\,2$

Strain due to initial prestress $\Big\} = \dfrac{115\cdot 5}{94\cdot 2} \times \dfrac{1}{200} = 0\cdot 006$

Strain developed at level of steel during decompression assuming a compression of 20 N/mm² at the level of the steel $\Bigg\} = \dfrac{20\times 6}{200\,000} = 0\cdot 0006$

Hence total strain at the level of the steel $\Big\} = 0\cdot 010\,28$

Strain in steel

$$\dfrac{0\cdot 80 f_{pu}}{1\cdot 15} = \dfrac{0\cdot 8\times 165}{94\cdot 2\times 1\cdot 15\times 200} = 0\cdot 006$$

Strain at $\dfrac{f_{pu}}{1\cdot 15} = 0\cdot 005 + \dfrac{165}{1\cdot 15\times 200\times 94\cdot 2}$

$$= 0\cdot 005 + 0\cdot 0076$$

$$= 0\cdot 0126$$

Hence stress in steel

$$= \frac{0 \cdot 80 f_{pu}}{\gamma_m} + \frac{0 \cdot 2 f_{pu}}{\gamma_m} \times \frac{0 \cdot 004}{0 \cdot 0066}$$

$$= \frac{0 \cdot 92 f_{pu}}{\gamma_m}$$

It will be assumed that the compression is uniform and equal to $0 \cdot 40 f_{cu}$ in the compression zone; the centre of compression is assumed at 0·42 of the neutral axis depth from the top. The top slab is replaced by the transformed area $= 0 \cdot 80 \times 150 \times 15 = 1800$. (The coefficient 0·80 accounts for the different moduli of elasticity of the concrete of the girder and the slab.)

$$\text{Compression in top slab} = \frac{1800 \times 0 \cdot 40 \times 40 \times 100}{1000}$$

$$= 2880 \text{ kN}$$

$$\left.\begin{array}{l}\text{Compression in top}\\ \text{flange of girder}\end{array}\right\} = 0 \cdot 40 \times \frac{40 \times 100}{1000} \times 60 \times 12 \cdot 5$$

$$= 1200 \text{ kN}$$

$$\text{Compression in web} = (95 - 27 \cdot 5) \times 14 \times \frac{40 \times 100}{1000} \times 0 \cdot 40$$

$$= 1512 \text{ kN}$$

$$\text{Compression in fillets} = 2 \times 0 \cdot 5 \times 23 \times 5 \times 0 \cdot 4 \times \frac{40 \times 100}{1000}$$

$$= 184 \text{ kN}$$

$C =$ Total compression $= 5776$ kN

$T =$ Total tension $= 0 \cdot 92 \times \dfrac{165}{1 \cdot 15} \times 42 = 5544$ kN

The total tension and compression are nearly equal. One more trial will lead to greater accuracy.

$C = T \approx 5600$ kN say.

$$\left.\begin{array}{l}\text{Ultimate moment of}\\ \text{resistance}\end{array}\right\} = 5600 \times \frac{(175 - 0 \cdot 42 \times 95)}{100}$$

$$= 5600 \times 1 \cdot 351$$

$$= 7565 \cdot 6 \text{ kN m} > 7193 \cdot 895 \text{ kN m}$$

Hence the structure is safe.

Design of post-tensioned members

(9) Checking at other sections

Checks are necessary at other sections to ensure that the permissible stresses are not exceeded at transfer and working loads. Ultimate flexural resistances at other sections have to be computed to make sure that they are greater than the bending moments occurring at those sections in the ultimate limit state for flexure. These checks are left to the reader as an exercise.

(10) Check for shear

V_u at support $= (1\cdot4$ D.L. shear $+ 1\cdot6$ L.L. shear)

Shear due to girder weight	$= 1\cdot4 \times 9\cdot77 \times 15$	$= 205\cdot17$ kN
Shear due to slab	$= 1\cdot4 \times 5\cdot40 \times 15$	$= 113\cdot40$ kN
Shear due to uniform live load	$= 1\cdot6 \times 18\cdot85 \times 15$	$= 452\cdot40$ kN
Shear due to knife edge load with load at support	$= 1\cdot6 \times 78\cdot26$	$= 125\cdot22$ kN
Shear due to asphalt and railing	$= 1\cdot4 \times 3 \times 15$	$= 63\cdot00$ kN
	Total	$= 959\cdot19$ kN

Maximum ultimate shear V_u at other sections are computed in a similar manner and the variation of V_u is plotted in Fig. 7.4.

Check for maximum shear stress

The effective depth of the beam will be assumed to be $195 - 13 = 182$ cm corresponding to the position of the lower-most prestressing tendons.

$$\text{Maximum shear stress} = \frac{959\cdot19 \times 1000}{182 \times 14 \times 100}$$

$$= 3\cdot76 \text{ N/mm}^2 < 4\cdot75 \text{ N/mm}^2$$

which is permissible.

Check at support section

Both V_{co} and V_{cr} have to be computed.

f_{cp} = compression at centroidal axis due to prestress

$$= \frac{3\,880\,000}{5872\cdot5 \times 100} = 6\cdot60 \text{ N/mm}^2$$

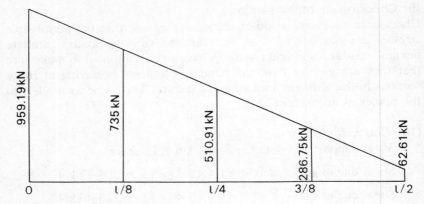

Fig. 7.4 Shear force diagram

$\dfrac{V_{co}}{bh}$ from Table $39 = 2 \cdot 10 + \dfrac{0 \cdot 6}{2 \cdot 0} \times 0 \cdot 20 = 2 \cdot 16 \text{ N/mm}^2$

$V_{co} = \dfrac{2 \cdot 16 \times 14 \times 195}{1000} = 589 \cdot 7 \text{ kN}$

In addition, the vertical components of the forces in the inclined tendons are to be added. From the parabolic profiles of cables shown in Fig. 7.5(a) their inclinations at the support section can be computed. For example, in Fig. 7.5(b) consider the profile of tendon 4. Measuring x from the left end, the equation of the tendon is

$$y = \dfrac{4 \times 100}{3000 \times 3000} x(1-x)$$

$$\left(\dfrac{dy}{dx}\right)_{x=0} = \dfrac{4 \times 100}{3000 \times 3000} \times 3000 = 0 \cdot 133$$

Sin $\theta \approx \tan \theta$

Hence vertical component $= 115 \cdot 5 \times 0 \cdot 80 \times 0 \cdot 133 \times 7$
$= 86 \cdot 02 \text{ kN}$

Tendon 5

Referring to Fig. 7.5, $\tan \theta = \dfrac{4 \times 113}{3000} = 0 \cdot 15$

Vertical component $= 115 \cdot 5 \times 0 \cdot 8 \times 0 \cdot 15 \times 7 = 97 \cdot 02 \text{ kN}$

Tendon 6

$\tan \theta = \dfrac{4 \times 126}{3000} = 0 \cdot 168$

Design of post-tensioned members 99

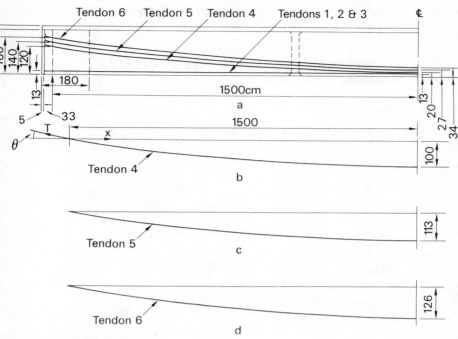

Fig. 7.5 Cable profiles

Vertical component = $115·5 \times 0·8 \times 0·16 \times 7 = 108·66$ kN
Total vertical component of all tendons   = $291·70$ kN
Hence V_{co} at support = $589·70 + 291·70$ = $881·4$ kN
$V_u = 959·19 > V_{co} = 881·4$ kN.

In any case nominal shear reinforcement is necessary. It may now be checked to see if V_u will still be greater than V_{co} when the value of the nominal stirrups is added.

Area of nominal reinforcement
$A_{sv} = 0·002 b_t s_v$
Let the spacing of stirrups be s_v and $A_{sv} = 0·002 \times 14 \times s_v$
Let two-legged 10 mm diameter stirrups with an area of $1·57$ cm^2 be provided so that $1·57 = 0·002 \times 14 \times s_v$
$s_v = 56$ cm.

Shear force carried by the stirrup $= \dfrac{1·57}{1·15} \times \dfrac{250\,000}{1000} \times \dfrac{182}{56}$
   $= 110·98$ kN.

Adding this to V_{co}, shear resistance of the section at the support is:

$V_{co} = 881\cdot40 + 110\cdot98 = 992\cdot38$
$V_{co} = 992\cdot38$ kN is now greater than $V_u = 959\cdot19$ kN.
Hence, the section is safe.

Check at l/8

Both V_{co} and V_{cr} are to be computed.

Check for V_{cr}

Position of tendon 4 above bottom fibre $= 120 - \dfrac{4 \times 100}{3000 \times 3000} \times 375 \ (2625)$

$= 120 - 43\cdot75 = 76\cdot25$ cm

Position of tendon 5 above bottom fibre $= 140 - 0\cdot4375 \times 113 = 90\cdot56$ cm

Position of tendon 6 above bottom fibre $= 160 - 0\cdot4375 \times 126 = 104\cdot87$ cm

Position of the resultant of the tendons above bottom fibre $= \dfrac{(3 \times 13 + 1 \times 76\cdot25 + 1 \times 90\cdot56 + 1 \times 104\cdot87)}{6}$

$= 51\cdot78$ cm

Eccentricity of prestressing force $= 86\cdot92 - 51\cdot78 = 35\cdot14$ cm

Prestress after losses at effective depth, i.e., at the level of the straight tendons $= \dfrac{42 \times 115\cdot5 \times 0\cdot8 \times 1000}{1000}$

$\times \left(\dfrac{1}{4072\cdot5} + \dfrac{35\cdot14 \times 73\cdot92}{17\,083\,792} \right)$

$= 15\cdot43$ N/mm^2

$M_0 = 0\cdot80 \times 15\cdot43 \times \dfrac{17\,083\,792}{73\cdot92} = 2852\cdot85$ kN m

Design of post-tensioned members 101

Ultimate shear V at $l/8 = 735$ kN.
Ultimate bending moment M at $l/8$ is computed as follows:

$$\left.\begin{array}{l}\text{Due to uniformly distributed}\\ \text{loads}\end{array}\right\} = 0\cdot4375 \times 5372\cdot775$$

$$= 2350\cdot589 \text{ kN m}$$

$$\left.\begin{array}{l}\text{Due to knife-edge load}\\ \text{(with load at section)}\end{array}\right\} = 78\cdot26 \times \frac{30}{8} \times \frac{7}{8} \times \frac{30}{30} \times 1\cdot6$$

$$= 410\cdot86499 \text{ kN m}$$

Total ultimate moment $= 2761\cdot4539$ kN m

$$\frac{V}{M} \times M_0 = \frac{2852\cdot85}{2761\cdot4539} \times 735 = 759\cdot32629$$

As this contribution itself is greater than 735, no further calculation is needed for checking V_{cr}.

V_{co} **at** $l/8$

The vertical components of tendons 4, 5 and 6 have to be computed at the section.

$$\left.\begin{array}{l}\text{Vertical component}\\ \text{of tendon 4}\end{array}\right\} = 7 \times 115\cdot5 \times 0\cdot8 \times 0\cdot10$$

$$= 64\cdot68 \text{ kN}$$

$$\left.\begin{array}{l}\text{Vertical component}\\ \text{of tendon 5}\end{array}\right\} = 7 \times 115\cdot5 \times 0\cdot8 \times 0\cdot113$$

$$= 73\cdot09 \text{ kN}$$

$$\left.\begin{array}{l}\text{Vertical component}\\ \text{of tendon 6}\end{array}\right\} = 7 \times 115\cdot5 \times 0\cdot8 \times 0\cdot126$$

$$= 81\cdot50 \text{ kN}$$

$$\left.\begin{array}{l}\text{Total vertical}\\ \text{component}\end{array}\right\} = 219\cdot27 \text{ kN}$$

102 Modern prestressed concrete design

V_{co} at the section including the shear capacity of nominal stirrups = $589 \cdot 70 + 219 \cdot 27 + 110 \cdot 98 = 919 \cdot 95$. This is higher than $V_u = 735$. Hence, the nominal shear steel provided is adequate.

Check at l/4

First V_{co} will be computed.

$$\text{Vertical component of tendon 4} \} = 7 \times 115 \cdot 5 \times 0 \cdot 8 \times \frac{4 \times 100 \times 1500}{3000 \times 3000}$$

$$= 7 \times 115 \cdot 5 \times 0 \cdot 8 \times 0 \cdot 0667$$

$$= 43 \cdot 12 \text{ kN}$$

$$\text{Vertical component of tendon 5} \} = 7 \times 115 \cdot 5 \times 0 \cdot 8 \times \frac{4 \times 113 \times 1500}{3000 \times 3000}$$

$$= 7 \times 115 \cdot 5 \times 0 \cdot 8 \times 0 \cdot 0753$$

$$= 48 \cdot 72 \text{ kN}$$

$$\text{Vertical component of tendon 6} \} = 7 \times 115 \cdot 5 \times 0 \cdot 8 \times \frac{4 \times 126}{3000 \times 3000} \times 1500$$

$$= 7 \times 115 \cdot 5 \times 0 \cdot 8 \times 0 \cdot 084$$

$$= 54 \cdot 33 \text{ kN}$$

$$\text{Total vertical component} \} = 146 \cdot 17 \text{ kN}$$

$V_{co} = 589 \cdot 70 + 146 \cdot 17 + 110 \cdot 98 = 846 \cdot 8512$

$V_u = 510 \cdot 91$

$V_{co} > V_u$

Check for V_{cr}

$$\text{Height of tendon 4 from bottom fibre} \} = 120 - \frac{4 \times 100}{3000 \times 3000} \times 2250 \times 750$$

$$= 45 \text{ cm}$$

$$\text{Height of tendon 5 above bottom fibre} \} = 140 - 0 \cdot 75 \times 113 = 55 \cdot 25 \text{ cm}$$

Design of post-tensioned members

$$\left.\begin{array}{l}\text{Height of tendon 6}\\ \text{above bottom fibre}\end{array}\right\} = 160 - 0{\cdot}75 \times 126 = 65{\cdot}5 \text{ cm}$$

$$\left.\begin{array}{l}\text{Height of tendon}\\ \text{resultant above}\\ \text{bottom fibre}\end{array}\right\} = \frac{3 \times 13 + 1 \times 45 + 1 \times 55{\cdot}25 + 1 \times 65{\cdot}5}{6}$$

$$= 34{\cdot}13 \text{ cm}$$

$$\left.\begin{array}{l}\text{Eccentricity of}\\ \text{prestressing force}\end{array}\right\} = 86{\cdot}92 - 34{\cdot}13 = 52{\cdot}79 \text{ cm}$$

$$\left.\begin{array}{l}\text{Prestress after}\\ \text{losses at the}\\ \text{effective depth}\\ \text{of section corres-}\\ \text{ponding to the}\\ \text{level of straight}\\ \text{tendons}\end{array}\right\} = \frac{42 \times 115{\cdot}5 \times 0{\cdot}8 \times 1000}{100}$$

$$\times \left(\frac{1}{4072{\cdot}5} + \frac{52{\cdot}79 \times 73{\cdot}92}{17\,083\,792}\right)$$

$$= 18{\cdot}39 \text{ N/mm}^2$$

$$M_0 = \frac{0{\cdot}80 \times 18{\cdot}39 \times 17\,083\,792}{73{\cdot}92} \times \frac{1}{1000} = 3400 \text{ kN m}$$

V at $l/4 = 510{\cdot}91$ kN

$$\frac{V}{M} \times M_0 = 3400 \times \frac{510{\cdot}91}{M}$$

The value of M, the ultimate bending moment at the section is computed as follows:

$$\left.\begin{array}{l}\text{B.M. at centre of span}\\ \text{due to uniformly dis-}\\ \text{tributed loads}\end{array}\right\} = 5372{\cdot}775 \text{ kN m}$$

$$\left.\begin{array}{l}\text{Corresponding moment}\\ \text{at } l/4\end{array}\right\} = 0{\cdot}75 \times 5372{\cdot}775 = 4029{\cdot}58 \text{ kN m}$$

$$\left.\begin{array}{l}\text{B.M. due to knife edge}\\ \text{load at the section}\end{array}\right\} = 1{\cdot}6 \times 78{\cdot}26 \times \frac{l}{4} \times \frac{3l}{4} \times \frac{1}{l}$$

$$= 1{\cdot}6 \times 78{\cdot}26 \times \frac{3}{16} \times 30$$

$$= 704{\cdot}34 \text{ kN m}$$

Total ultimate moment at section $= 6077 \cdot 135$ kN m

$\dfrac{V}{M} \times M_0 = 3400 \times \dfrac{510 \cdot 91}{6077 \cdot 135} = 285 \cdot 84$ kN

$f_{pe} = 0 \cdot 80 \times 0 \cdot 70 \times f_{pu} = 0 \cdot 56 f_{pu}$

$(1 - 0 \cdot 55 \times 0 \cdot 56) = 0.692$

$0 \cdot 692 v_c bd = 0 \cdot 692 \times 0 \cdot 80 \times 14 (195 - 34 \cdot 125) \times \dfrac{100}{1000}$

$\qquad\qquad\qquad\qquad = 124 \cdot 685$ kN

$V_{cr} = 285 \cdot 84 + 124 \cdot 69 + 110 \cdot 98 = 638 \cdot 581$ kN

$V_u = 510 \cdot 91$ kN $\quad V_{cr} > V_u$

Hence the nominal stirrups provided are enough.

Check at $\tfrac{3}{8} l$

Height of tendon 4 from bottom fibre $= 120 - \dfrac{4 \times 100}{3000 \times 3000} \times 1125 \times 1875$

$\qquad\qquad\qquad = 120 - 93 \cdot 75 = 26 \cdot 25$ cm

Height of tendon 5 from bottom fibre $= 140 - \dfrac{4 \times 113}{3000 \times 3000} \times 1125 \times 1875$

$\qquad\qquad\qquad = 34 \cdot 06$ cm

Height of tendon 6 above bottom fibre $= 160 - \dfrac{4 \times 126}{3000 \times 3000} \times 1125 \times 1875$

$\qquad\qquad\qquad = 41 \cdot 875$ cm

Height of resultant of tendon from bottom fibre $= \dfrac{3 \times 13 + 26 \cdot 25 + 34 \cdot 06 + 41 \cdot 88}{6}$

$\qquad\qquad\qquad = \dfrac{39 + 26 \cdot 25 + 34 \cdot 06 + 41 \cdot 88}{6}$

$\qquad\qquad\qquad = 23 \cdot 53$ cm

Eccentricity $\qquad = 86 \cdot 92 - 23 \cdot 53 = 63 \cdot 39$ cm

Design of post-tensioned members

Prestress at the effective depth corresponding to level of straight tendons

$$= \frac{42 \times 115 \cdot 5 \times 0 \cdot 8 \times 1000}{100}$$

$$\times \left(\frac{1}{4072 \cdot 5} + \frac{63 \cdot 39 \times 73 \cdot 92}{17\,083\,792} \right)$$

$$= 20 \cdot 08 \text{ N/mm}^2$$

$$M_0 = 0 \cdot 80 \times 20 \cdot 08 \times \frac{17\,083\,792}{73 \cdot 92} = 3712 \cdot 45 \text{ kN m}$$

The ultimate bending moment M at the section is computed as follows:

B.M. due to uniformly distributed loads $= \tfrac{15}{16} \times 5372 \cdot 775$

$\qquad = 5036 \cdot 98 \text{ kN m}$

B.M. due to knife edge load at section $= \tfrac{3}{8} \times \tfrac{5}{8} l \times 78 \cdot 26 \times 1 \cdot 6$

$\qquad = 880 \cdot 42 \text{ kN m}$

Ultimate B.M. at section $= 5917 \cdot 4049 \text{ kN m}$

$$\frac{V}{M} \times M_0 = 286 \cdot 75 \times \frac{3712 \cdot 45}{5917 \cdot 40} = 179 \cdot 90 \text{ kN}$$

$$0 \cdot 692 v_c bd = 0 \cdot 692 \times 0 \cdot 80 \times 14(195 - 23 \cdot 53) \times \frac{100}{1000}$$

$\qquad = 132 \cdot 89 \text{ kN}$

Total V_{cr} $= 312 \cdot 79 \text{ kN}$

V_u $= 286 \cdot 75 \text{ kN}$

$V_{cr} > V_u$ even without considering the contribution of nominal shear reinforcement provided.
Next V_{co} has to be computed.

Vertical component of tendon 4 $= 7 \times 115 \cdot 5 \times 0 \cdot 8 \times \dfrac{400 \times 750}{3000 \times 3000}$

$\qquad = 21 \cdot 56 \text{ kN}$

$$\left.\begin{array}{l}\text{Vertical component of}\\ \text{tendon 5}\end{array}\right\} = 7 \times 115 \cdot 5 \times 0 \cdot 8 \times \frac{4 \times 113 \times 750}{3000 \times 3000}$$

$$= 24 \cdot 35 \text{ kN}$$

$$\left.\begin{array}{l}\text{Vertical component of}\\ \text{tendon 6}\end{array}\right\} = 7 \times 115 \cdot 5 \times 0 \cdot 8 \times \frac{4 \times 126 \times 750}{3000 \times 3000}$$

$$= 27 \cdot 17 \text{ kN}$$

$$\left.\begin{array}{l}\text{Total vertical}\\ \text{component}\end{array}\right\} = 73 \cdot 08 \text{ kN}$$

$V_{co} = 589 \cdot 70 + 73 \cdot 08 + 110 \cdot 98 = 773 \cdot 76$ kN

$V_u = 286 \cdot 75$

Hence $V_{co} > V_u$.

Check at $\frac{l}{2}$

The Code requires that a minimum value of $V_{cr} = 0 \cdot 1\sqrt{f_{cu}} \times bd$ is to be assumed.

$$\text{Minimum } V_{cr} = 0 \cdot 10 \times \sqrt{40} \times 14 \times 182 \times \tfrac{100}{1000} = 161 \cdot 1 \text{ kN}$$

$V_u = 62 \cdot 61$ kN $V_{cr} > V_u$

V_{co} is not applicable as the section would definitely have cracked in flexure.

(11) Check for shear at interface

Total vertical shear at support due to design loads:

Due to girder weight	$= 9 \cdot 77 \times 15$	$= 146 \cdot 55$
Due to slab	$= 5 \cdot 40 \times 15$	$= 81 \cdot 00$ kN
Due to uniformly distributed load	$= 18 \cdot 85 \times 15$	$= 282 \cdot 75$ kN
Due to knife edge load	$=$	$78 \cdot 26$ kN
Due to asphalt and railing	$= 3 \times 15$	$45 \cdot 00$ kN

$$\text{Total } V = 633 \cdot 50 \text{ kN}$$

First moment of area of part above interface $= Q$

$$= 150 \times 15 \times (77 \cdot 25 - 7 \cdot 5)$$
$$= 156\,937 \cdot 5 \text{ cm}^3$$

Moment of inertia of composite section

$$= 26\,895\,875 \text{ cm}^4$$

Width of surface at interface b_e

$$= 60 \text{ cm}$$

Horizontal shear at interface

$$= \frac{VQ}{Ib}$$

$$= \frac{633 \times 1000 \times 156\,937 \cdot 5}{26\,895\,875 \times 60} \times \frac{1}{100}$$

$$= 0 \cdot 61613 \text{ N/mm}^2$$

The horizontal shear stress is within the permissible limit of $1 \cdot 32$ N/mm² corresponding to Type 3 surface as defined in CP 110 Clause 5.4.3.4. This presupposes that the surface receiving the *in situ* concrete is brushed with a stiff brush and links are provided. The second requirement is met by extending the nominal two-limbed stirrups into the *in situ* slab.

7.3 Design of End Blocks

The stress distribution in the neighbourhood of end anchorages transferring concentrated loads is best understood by considering the case of a single anchorage transferring a concentrated load of P. Analytical [7.4], [7.5], [7.6], [7.7] and experimental studies [7.8], [7.9] and [7.10] show that in the lead-in zone extending to a distance of roughly $2a$ from the end, bursting tensile stresses develop in the two directions at right angles to the line of action of the load. The transverse stress distribution is of the form indicated in Fig. 7.6(a). Experiments [7.8] show that the stress changes from compression to tension at a distance of about $0 \cdot 2a$ from the end; the maximum bursting tensile stress occurs at $0 \cdot 5a$; the bursting tensile stress becomes negligible at a distance of $2a$ (Fig. 7.6(b)). The magnitude of the bursting tensile stress may conveniently be expressed as a fraction of the uniform compression p. (Fig. 7.6(a)). This fraction depends on a_1/a or b_1/b as the case may be. Two useful graphs Figs. 7.7 and 7.8 reproduced from reference [7.8] give the maximum tensile stress and the resultant tensile force respectively. In addition to the bursting zone, there is also a spalling zone close to the end where tensile stresses indicating the

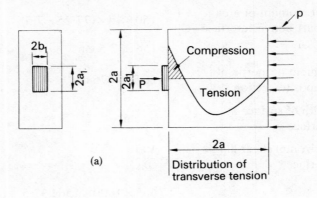

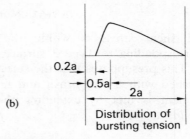

Fig. 7.6 Stresses in end block

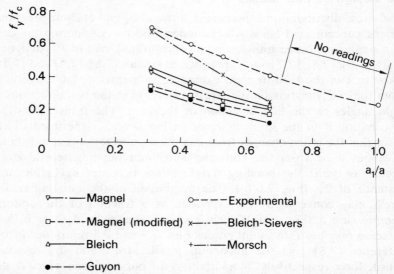

Fig. 7.7 Maximum transverse stress as ratio of uniform compression

Design of post-tensioned members 109

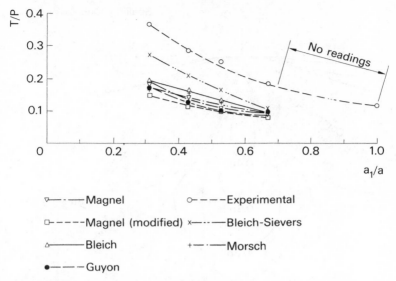

Fig. 7.8 Relation between tensile force T and a_1/a ratio

need for a reinforcing mesh very close to the edge. The bursting and spalling zones may be seen in Fig. 7.9.

Comprehensive experimental studies reported by the Cement and Concrete Association [7.8] and [7.9] have shown that the analytical method due to Guyon [7.5] underestimates the bursting tension.

A practical procedure for designing reinforcement of end blocks will be to utilize the method of successive resultants developed by Guyon and make use of the experimental results to arrive at the magnitude and position of the bursting tensile force.

The method of successive resultants essentially reduces the problem of an end block with a number of anchorages transferring concentrated forces to the problem of a series of symmetrical prisms each of which carries a single central concentrated force which may be the total resultant, group resultant or the individual force distributed over a length $2a_1$ in one direction and $2b_1$ in the other. Each prism will have a depth of $2a_1$, a being taken as the distance of the line of action of the force acting on the prism from the nearer free edge. This procedure implies that the bursting forces are critical along the lines of action of the total resultant of the forces, the lines of action of the resultants of groups of forces and finally along the lines of action of the individual forces. The procedure is best understood by going through Design Example 7.2.

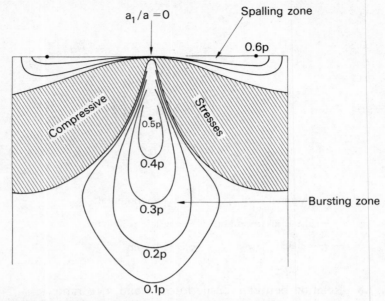

Fig. 7.9 Spalling and bursting zones (After Guyon)

Design Example 7.2

The end block considered is that of the post-tensioned beam of Design Example 7.1.

The prestressing forces at the ends being nearly horizontal, their vertical components may be ignored. The forces acting on the end block are shown in Fig. 7.10.

Forces before losses occur are considered. They are

Force in tendons 1, 2 and 3 = 3×165 = 495 kN
Force in tendon 4   =   = 165 kN
Force in tendon 5   =   = 165 kN
Force in tendon 6   =   = 165 kN

The resultant of all cables will be at a distance above the bottom face = $\dfrac{(495 \times 13)+(165 \times 3 \times 140)}{990} = 76\cdot5$ cm. Hence, $2a = 153$ cm. Size of anchorage = 152×152 mm.

Analysis will be made by the method of successive resultants. Computations will be made along the lines of action R (the resultant of all of the six cables), R_1, the resultant of tendons 1, 2 and 3 and R_3 the resultant of tendons 4, 5 and 6. It is also necessary to make such calculations along the lines of action of the individual cables 4, 5 and 6.

Design of post-tensioned members 111

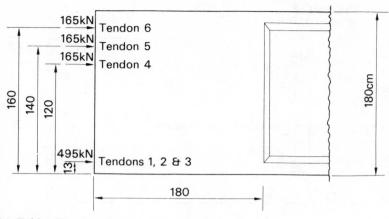

Fig. 7.10 Tendon forces on end block

These are omitted here as no new principles are involved. Although the method of successive resultants developed by Guyon will be employed, the higher experimental values reported by Rowe will be made use of in arriving at the reinforcement.

Along line of action of R = 990 kN

$2a_1 = 4 \times 152 = 608$ mm

$2a = 2 \times 765 = 1530$ mm

$\dfrac{a_1}{a} = 0.397$ or say 0.40

Resultant of tensile force from Fig. 7·8 as

$= 0.30 \times 990$

$= 297$ kN

Maximum $f_y = 0.65 \times \left(\dfrac{990\,000}{600 \times 1530} \right)$

$= 0.70$ N/mm^2

Experiments reported by Rowe show that this maximum tension will occur at $0.50a = 0.50 \times 76.5 = 38.25$ cm from the end. It is also known from experiments that the zero value of f_y occurs at approximately $0.2a$ from the end, i.e., $0.2 \times 76.5 = 15.3$ cm from the end. Approximating the curvilinear distribution by a triangle, we have the following distribution (Fig. 7.12). The tensile bursting force to be provided for is 297 kN. The Cement and Concrete Association's *Research Report* No.

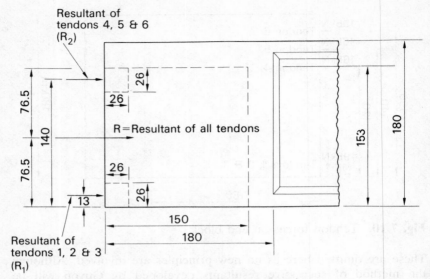

Fig. 7.11 Successive resultant of symmetric prisms

9 [7·8] suggests that reinforcement need be provided only for the residual tensile force after deducting the load that the concrete of the end block can carry in tension. Because of the various uncertainties involved, the tensile strength of the concrete is ignored and steel provided to carry the tension of 297 kN. Using 10 mm diameter bars, the number of bars required is $(297\,000 \times 1\cdot15)/(250 \times 78\cdot54) = 17\cdot39$ or say 18 bars. These will be uniformly distributed across the width of the block at a distance of 38·25 cm from the end. A similar calculation has to be done for steel to be provided in the other direction. It is conservative to provide the same steel at right angles without making an additional calculation.

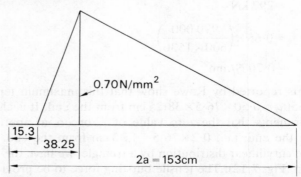

Fig. 7.12 Distribution of bursting tension along line of action of R

Along resultant R_1

Force $R_1 = 3 \times 165 = 495$ kN

$2a_1 = 152$ mm

$2a = 260$ mm

$\dfrac{a_1}{a} = \dfrac{152}{260} = 0.5846$

Uniform compression $= \dfrac{495\,000}{600 \times 260} = 0.318$

Maximum f_y $= 0.50 \times 0.318$

$= 0.159$ N/mm^2 at 0.5×13 cm $= 6.5$ cm

Resultant tensile force $= 0.21 \times 495 = 104$ kN

Number of bars $= \dfrac{104\,000 \times 1.15}{250 \times 78.54} = 6.10$ or 7

Along resultant R_2
Calculations will be the same as above.

Along line of action of cable 4

$2a_1 = 152$

$2a = 660$

$\dfrac{a_1}{a} = \dfrac{152}{660} = 0.23$

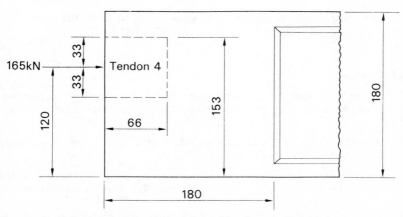

Fig. 7.13 Resultant of symmetric prism for tendon 4

114 Modern prestressed concrete design

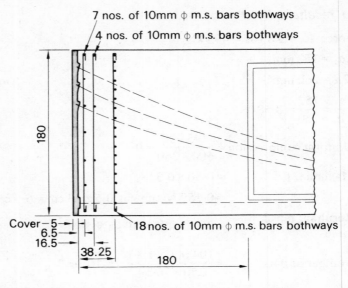

Fig. 7.14 Scheme of reinforcement in the end block

Experimental results are not available for $a_1/a = 0 \cdot 23$.
Extrapolating the curve in Fig. 7.8, the value of $T/P = 0 \cdot 40$ approximately.
Hence the bursting tensile force $= 0 \cdot 40 \times 165$ kN $= 66$ kN

$$\left.\begin{array}{l}\text{Hence number of 10 mm bars}\\ \text{required}\end{array}\right\} = \frac{66000 \times 1 \cdot 15}{78 \cdot 54 \times 250}$$

$$= 3 \cdot 92 \text{ or say 4 bars}$$

The reinforcement will be placed at $16 \cdot 5$ cm from the edge.
The scheme of reinforcing is shown in Fig. 7.14.

References

[7.1] Rahman, Abdul, P. M., and Karim, Abdul, E., 'Bonding precast concrete elements using epoxy resins', *Indian Concrete Journal*, May 1975, pp. 142–7.

[7.2] Mustafa, Saad, E., 'Ultimate load test of a segmentally constructed concrete I beam', *PCI Journal*, July/August, 1974.

[7.3] Ernani, Diaz, B., 'The technique of glueing precast elements of the Rio-Niteroi Bridge', *RILEM, Materials and Structures*, January/February 1975.

[7.4] Magnel, G., *Prestressed Concrete*, Concrete Publications, 1950.

[7.5] Guyon, Y., 'Prestressed concrete', *Contractors' Record*, 1960.

[7.6] Sievers, H., 'Über den Spannungszustand im Bereich der Ankerplatten von Spanngliedern vorgespannter Stahlbeton-konstruktionen', *Der Bauingenieur*, Vol. 31, No. 4, April 1956, pp. 134–5.

[7.7] Ramaswamy, G. S., and Goel, Harish, 'Stresses in end blocks of prestressed beams by lattice-analogy', *Proceedings of World Conference on Prestressed Concrete*, San Francisco, 1957.

[7.8] Zielinski, J., and Rowe, R. E., 'An investigation of the stress distribution in the anchorage zones of post-tensioned concrete members', Cement and Concrete Association, *Research Report No. 9*, 1960.

[7.9] Zielinski, J., and Rowe, R. E., 'The stress distribution associated with groups of anchorages in post-tensioned concrete members', Cement and Concrete Association, *Research Report No. 13*, 1962.

[7.10] Chirstodoulides, S. P., 'Three-dimensional investigation of stresses in the end anchorage blocks of a prestressed gantry beam', *The Structural Engineer*, Vol. 35, No. 9, September 1957, pp. 349–56.

8
Transition from fully prestressed to reinforced concrete

The transition from fully prestressed (Class 1) to reinforced concrete (Class 4) structures takes place through intermediate degrees of prestress represented by Class 2 and Class 3 structures. It can also happen that in the same structure different regions may conform to Class 1, Class 2 or Class 3 conditions.

8.1 Historical Notes

Only 'full prestressing' was being practised when prestressed concrete was first introduced. Full prestressing demands that no *flexural tensile stresses* occur under working loads. *Principal tensile stresses* cannot, however, be eliminated unless the member is also provided with tensioned stirrups.

In the early days of prestressed concrete, Freyssinet advocated full prestressing. Recognizing no half-way house between fully prestressed and reinforced concrete, he maintained that each intermediate solution was less satisfactory than either of them. That he changed his mind in later years is evident from the following extract from his speech on the occasion of his jubilee as an engineer [8.1].

> 'I take the common example of a broad highway bridge. The regulation requires that under the most unfavourable conditions, there shall remain a residual compression of 8 kg/cm^2. In my first prestressed concrete bridge, I imposed on myself residual compression even larger than this. They are no longer necessary now that the art of calculating and constructing bridges has made such progress. It may be admitted that in railroad bridges subjected to live loads approaching the maximum several thousand times a day, tension should be eliminated in concrete even though this means asking of prestressed concrete more than what is asked of other materials.... In a road bridge where there is no chance of one in ten thousand of the unfavourable load being applied twice in the life of the structure, there is no disadvantage in allowing tensions of the order of 50 kg/cm^2 in a concrete whose strains are appropriately guided; to forbid it is a pure waste of public money. And to

Transition from fully prestressed to reinforced concrete

what purpose? In practice, structures subject to these Draconian conditions find themselves in competition with structures reinforced with high yield strength deformed bars.... Regulations of an exaggerated severity lead to the substitution of a less good structure in the place of a better one...'.

The case for intermediate degrees of prestress cannot be more eloquently stated.

A survey of relative cost figures of prestressing steel and high yield strength high bond deformed bars in a number of countries shows that prestressing steel including sheathing, stressing, anchorages and grouting costs 3·5 to 4·0 times as much as high yield strength deformed bars placed in position. These figures are confirmed by the price ratio of 3 to 5 reported for Switzerland by Thürlimann and Calfisch [8.2]. This price ratio is the most compelling reason for adopting limited or partial prestressing. Saving in prestressing steel so realized leads to an overall saving in cost. Thürlimann estimates that the saving in prestressing steel will be as large as 30%, if partial prestressed concrete structures are designed according to the Swiss Code SIA 162 of 1968 [8.3]. This is confirmed by Ramaswamy and Raman [8.4]. Fallacious comparisons are often made between fully prestressed concrete and reinforced concrete alternatives. Let us suppose that for the same job, two competitive bids are offered—one of them of fully prestressed concrete and the other of reinforced concrete. The fully prestressed design is crack-free under design loads. On the other hand, the reinforced concrete alternative is deemed to be cracked below the neutral axis from the very beginning. In such instances, we are obviously not comparing two comparable alternatives. The comparison will be somewhat fairer to prestressed concrete, if partial prestressing is adopted. It is this line of thinking that has led to the current trend towards Class 2 and Class 3 structures.

Conceptually, Class 3 members may be regarded either as improved reinforced concrete or partially prestressed concrete structures. In the former case, they will be designed essentially as reinforced concrete members with just enough prestress introduced to ensure satisfactory serviceability (control of cracking and deflexion within desired limits). In the latter, they will be designed as prestressed concrete members, the areas of tensioned and untensioned steel provided in them being dictated respectively by the serviceability and ultimate limit state requirements. Historically, partial prestressing appears to have originated with Emperger in 1939 who was merely trying to improve the serviceability of Class 4 members reinforced with medium tensile deformed bars by introducing a few prestressed wires [8.5]. Abeles who was closely associated with Emperger's experiments developed

118 Modern prestressed concrete design

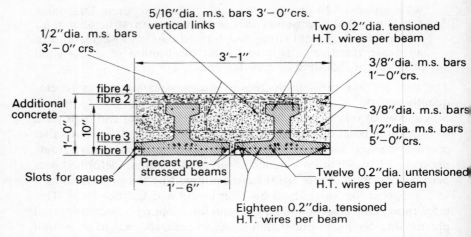

Fig. 8.1 Typical partially prestressed bridge girders standardized by British Railways, Eastern Region

these ideas further in 1940–42 and suggested that instead of medium tensile deformed bars, prestressing steel itself be used as untensioned reinforcement. Alternatively, he recommended that the entire prestressing steel required to meet the ultimate limit state requirements be stressed to a lesser extent than is normally done in full prestressing. The ideas advanced by Abeles were accepted by the Chief Engineer's department, British Railways, Eastern Region and starting from 1948 a number of bridges of composite construction (Fig. 8.1) were built in the span range of 9 to 15 m, permitting a tensile stress in the homogeneous section of $3 \cdot 85$ N/mm^2. This was subsequently raised to $4 \cdot 55$ N/mm^2 when tests revealed no visible cracks on repeating the loading cycle a million times and varying the stress from 0 to $4 \cdot 55$ N/mm^2.

As already noted, Abeles recommended the use of prestressing steel itself as untensioned reinforcement. Current practice is to use mild steel, high yield strength deformed bars or prestressing steel as passive (untensioned) reinforcement. High yield strength deformed bars are to be preferred because they are less expensive than prestressing steel and have higher yield strengths and better crack control characteristics compared to mild steel.

The West German Code DIN 4227 had accorded recognition to partial prestressing concepts even before the CEB-FIP recommendations were introduced. The terminology used in German practice may be clarified by the following extract from the book by Leonhardt [8.6].

Transition from fully prestressed to reinforced concrete

'In full prestressing only small tensile stresses are exceptionally permitted under working load. In limited prestressing, tensile stresses up to 80% of the tensile strength are permitted in the uncracked condition (homogeneous section). "Moderate prestressing" comprises the range between ordinary reinforced concrete and limited prestressing.'

Now that CEB-FIP recommendations have evolved a systematic classification to designate intermediate degrees of prestressing, the terms 'partial prestressing', 'moderate' and 'limited' prestressing are best avoided.

8.2 Advantages of Class 2 and Class 3 Structures

The advantages of intermediate degrees of prestressing may be summed up as follows.

(1) Saving in prestressing steel and consequent saving in cost.

(2) Fully prestressed structures often cause camber problems. The full live loads specified in the codes seldom occur, and consequently the upward deflexion caused by prestress is only partially nullified by the live loads. The residual camber continues to increase with time on account of creep. Class 2 and Class 3 structures do not create such problems because the prestress is much less.

(3) Reference to Fig. 8.2 in which the load-deflexion characteristics of beams with different degrees of prestress are compared will

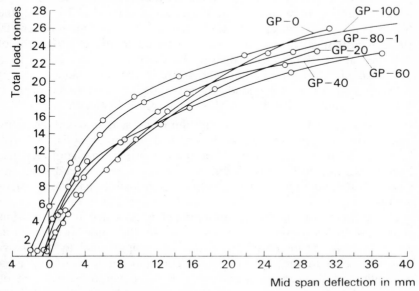

Fig. 8.2 Load deflection curves

show that the fully prestressed beam because of its higher rigidity is capable of less energy absorption (represented by the area under the load-deflexion curve). Class 2, Class 3 and Class 4 beams are seen to be more ductile.

(4) Less expensive medium tensile-strength deformed bars can be conveniently used for prestressing Class 3 members by employing the electro-thermal technique, because the prestress to be imparted is much less than for a Class 1 member.

8.3 Design Principles of Class 2 Structures

The design principles of Class 2 structures are not substantially different from those of Class 1 structures because under working loads, the section is not cracked and the properties of the homogeneous (uncracked) section can be used in the analysis just in the same manner as for Class 1 structures.

8.4 Design Principles of Class 3 Structures

The design of Class 3 members involves the computation of crack-widths under working loads and restricting them to the values prescribed in the Codes. There are two different methods which are commonly employed; several variations of each are also in use.

Method 1: In this method, the hypothetical tensile stress in the extreme tension-fibre under working loads is computed by using the homogeneous (uncracked) section. Based on experimental results, these hypothetical tensile stresses are related to the crack-widths of 0.10 mm and 0.20 mm prescribed in the Codes.

Method 2: In this method, the stress in the untensioned steel is computed under working loads by using the cracked section and is related to crack-widths by using formulae given in the Codes.

Method 1 has the merit of extreme simplicity as it avoids the lengthy calculations involved in making the cracked-section analysis. It is also found to be quite dependable. For these reasons, this approach has been prescribed in CP 110. Table 8.1, which is reproduced from the Code, relates nominal tensile stresses to permissible maximum crack-widths.

If additional untensioned reinforcement is provided in the tension zone close to the tension face of the concrete, the values prescribed in Table 8.1 may be increased by an amount proportional to the cross-sectional areas of the reinforcement. This amount is expressed as a percentage of the cross-sectional area of the concrete at the rate of 4 N/mm^2 for members of groups A and B and 3 N/mm^2 for members

Table 8.1 Hypothetical tensile stresses for Class 3 members

	Limiting crack-width mm	Stress in N/mm², for concrete grade		
		30	40	50 and over
A. Pretensioned tendons	0·10	—	4·1	4·8
	0·20	—	5·0	5·8
B. Grouted post-tensioned tendons	0·10	3·2	4·1	4·8
	0·20	3·8	5·0	5·8
C. Pretensioned tendons distributed in the tensile zone and positioned close to the tension face of the concrete	0·10	—	5·3	6·3
	0·20	—	6·3	7·3

The values given in Table 8.1 should be multiplied by the coefficients given in Table 8.2 to take account of the depth factor.

of group C for each one per cent of additional reinforcement, subject to the condition that in no case shall the hypothetical tensile stress exceed one quarter of the characteristic cube strength of the concrete. Bennett and Chandrasekhar [8.7] have proposed a formula linking nominal tensile stresses and crack-widths.

Method 2, being more rigorous, was preferred for inclusion in the CEB–FIP recommendations which give the following formulae linking the crack-width with the stress in the untensioned steel under working loads:

$$w_{max} = [f_{s(u)} - 4000] \times 10^{-6}$$ for non-repetitive loads and

and

$$w_{max} = f_{s(u)} \times 10^{-6}$$ for repetitive loads,

where w_{max} is the maximum crack-width in centimetres and $f_{s(u)}$ is the stress in the untensioned steel at working loads in N/cm². The application of Method 2 will involve the following steps:

(1) Choose the bending moment at which decompression is to take place (i.e., the stress in the extreme concrete fibre just becomes

Table 8.2 Depth factor for tensile stresses for Class 3 members

Depth of member (mm)	200 and under	400	600	800	1000 and over
Factor	1·10	1·0	0·90	0·80	0·70

122 Modern prestressed concrete design

zero). Normally, one may specify that decompression occurs under dead load. However, it has been pointed out by Walther and Bhal [8.8] that in order to realize fully the economy of Class 3 structures, decompression should take place even earlier and that under dead load, the stress in the extreme fibre should reach the tensile strength of the concrete. The area of prestressed steel A_{ps} is arrived at on the basis of the selected decompression bending moment.

(2) The area of prestressed steel so selected will be inadequate to meet the requirements of the ultimate limit state. The area of the untensioned steel $A_{s(u)}$ needed to supplement the resisting moment is found from the external bending moment to which the section is subjected at the ultimate limit state.

(3) A cracked-section analysis is carried out under the working-load bending moment M_T to determine the stress in the untensioned steel.

(4) The crack-width is computed from the stress in the untensioned steel and checked to see if it is within the prescribed limit.

(5) The design is revised if necessary.

8.5 Cracked-section Analysis

The cracked-section analysis under the working-load bending moment M_T is carried out as follows [8.9].

Consider the cross-section of a Class 3 beam submitted to the action of working load bending moment M_T (Fig. 8.3). The following assumptions are made:

(1) Strain distribution across the cross-section is linear.
(2) The effective prestressing force P_e in the stressed tendons

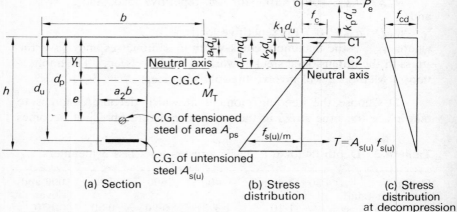

(a) Section   (b) Stress distribution   (c) Stress distribution at decompression

Fig. 8.3 Section of a class 3 beam

(after losses) does not appreciably change under the action of M_T and is hence assumed to be constant.

(3) The tensile strength of the concrete below the neutral axis is neglected.

(4) The influence of uncracked sections between cracks is ignored.

The centre of pressure which is originally coincident with the tendon when no external loads are acting will shift to the point O under the action of M_T. The total shift $= e + y_t + k_p d_u$. Hence, we may write

$$M_T = P_e(e + y_t + k_p d_u) \tag{8.1}$$

Or

$$k_p = \frac{M_T - P_e(e + y_t)}{P_e d_u} \tag{8.2}$$

The total compressions in the top flange and the part of the web below the flange are C_1 and C_2 respectively and their points of application are $k_1 d_u$ and $k_2 d_u$ from the top fibre. The following are easily verified

$$C_1 = \frac{f_c b d_u}{2} \left(\frac{2n - a_1}{n} \right) a_1 \tag{8.3}$$

$$C_2 = \frac{f_c b d_u}{2} \left[\frac{(n - a_1)^2}{n} \right] a_2 \tag{8.4}$$

$$k_1 = \left(\frac{3n - 2a_1}{2n - a_1} \right) \frac{a_1}{3} \tag{8.5}$$

$$k_2 = \left(\frac{2a_1 + n}{3} \right) \tag{8.6}$$

$$T = f_{s(u)} A_{s(u)} = \frac{m(1 - n)}{n} f_c p_u b d_u \tag{8.7}$$

where

$$p_u = \frac{A_{s(u)}}{b d_u} \text{ and } m \text{ is the modular ratio.}$$

Taking moments of forces about O, the equation of equilibrium $\Sigma M = 0$ may be written as

$$C_1(k_p + k_1) + C_2(k_p + k_2) - T(k_p + 1) = 0 \tag{8.8}$$

Summing up all forces in the plane algebraically and equating them to zero, i.e., $\Sigma F = 0$, we arrive at

$$C_1 + C_2 - T - P_e = 0 \tag{8.9}$$

Substituting for C_1, C_2, k_1, k_2 and T in equation (8.8) from equations (8.3), (8.4), (8.5), (8.6) and (8.7), we arrive at the following cubic

124 Modern prestressed concrete design

equation in n

$$n^3 + 3k_p n^2 + \left(\frac{K_1+K_2}{a_2} - K_1\right)n + \left(K_3 - \frac{K_3+K_2}{a_2}\right) = 0 \qquad (8.10)$$

where

$$K_1 = 3a_1(a_1 + 2k_p) \qquad (8.11)$$
$$K_2 = 6mp_u(1 + k_p) \qquad (8.12)$$
$$K_3 = a_1^2(2a_1 + 3k_p) \qquad (8.13)$$

Again, by substituting for C_1, C_2 and T in equation (8.9), we get

$$f_c = \frac{P_e}{bd_u}\left[\frac{2n}{a_1(2n-a_1) + a_2(n-a_1)^2 - 2mp_u(1-n)}\right] \qquad (8.14)$$

After solving for n from (8.10), f_c may be found from (8.14). When f_c is known, $f_{s(u)}$ is found from (8.7). If assumption (2) is not made, instead of the cubic equation as given in (8.10), we shall have a fifth degree equation in n when the neutral axis is outside the flange [8.10]. In the special case, when the neutral axis is within the flange, the fifth degree equation will degenerate into a cubic. The stress in the untensioned steel can also be found by an iterative graphical method due to Levi described by Guyon [8.11].

8.6 Design Examples

Computation of crack-widths using Methods 1 and 2 are best explained by means of examples.

Example 8.1

Problem: Design a Class 3 pretensioned girder of the cross-section shown in Fig. 8.4 for a span of 20 m to carry a live load of 9 kN/m. The crack-width under working loads is restricted to 0·20 mm. The concrete is to have a characteristic cube strength of 50 N/mm^2 and a cube strength at transfer of 40 N/mm^2. Losses in prestress may be estimated at 20% of the initial prestress. The tensioned steel is to consist of strands of 12·5 mm nominal diameter with a characteristic strength of 165 kN, the nominal area of cross-section being 94·2 mm^2. The untensioned steel shall be of cold-worked deformed bars with a characteristic strength of 425 N/mm^2.

Sectional properties

$$A_c = (100 \times 12) + (60 \times 30) = 3000 \text{ cm}^2$$

$$y_t = \frac{(1200 \times 6) + (60 \times 30 \times 42)}{1200 + 1800} = 27\cdot6 \text{ cm}$$

Transition from fully prestressed to reinforced concrete

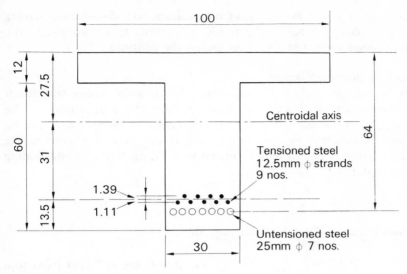

Fig. 8.4 Cross section of girder

Moment of inertia $I = 1\,487\,400$ cm^4

$$Z_t = \frac{1\,487\,400}{27 \cdot 6} = 54\,000 \text{ cm}^3$$

$$Z_b = \frac{1\,487\,400}{44 \cdot 4} = 33\,500 \text{ cm}^3$$

Bottom kern distance $= \dfrac{Z_t}{A_c} = 18$ cm

Girder weight $= \dfrac{0 \cdot 3000 \times 23000}{1000} = 6 \cdot 9$ or say $7 \cdot 0$ kN/m

Bending moments

Bending moment M_G due to self-weight of girder $\Big\} = \dfrac{7 \times 20 \times 20}{8}$

$= 350$ kN m

Bending moment M_L due to live load $\Big\} = \dfrac{9 \times 20 \times 20}{8}$

$= 450$ kN m

Total working load bending moment M_T $\Big\} = 350 + 450 = 800$ kN m

126 Modern prestressed concrete design

The value of e may be selected from practical considerations of having enough space left below the tensioned strands to accommodate the untensioned steel. Let $e = 31$ cm below the centroid.

State of decompression
Let us prescribe that decompression is to occur under the girder bending moment of M_G. At decompression, the distribution will be triangular as shown in Fig. 8.3(c). The centre of compression can be found by determining the compressive forces F_1 in the overhanging flange (70 cm × 12 cm) and F_2 in the web (72 cm × 30 cm) and finding the resultant of F_1 and F_2.

$$F_1 = 70 \times 12 \times \frac{f_{cd}}{2}\left(1 + \frac{5}{6}\right) = 770 f_{cd}$$

It will act at 5·82 cm from the top fibre.

$$F_2 = 72 \times 30 \times \frac{1}{2} \times f_{cd} = 1080 f_{cd}$$ and it will act at 24 cm from top.

The resultant of F_1 and F_2 will act at $(770 \times 5 \cdot 82 + 1080 \times 24)/1850 = 16$ cm from the top. Hence shift in the centre of pressure $= 58 \cdot 6 - 16 = 42 \cdot 6$ cm. Decompression bending moment $= P_e \times 42 \cdot 6$, where P_e is the effective prestress after losses.
Hence

$$42 \cdot 6 \, P_e = 35000$$

$$P_e = 830 \text{ kN.}$$

Initial prestressing force P (assuming 20% losses) $\Big\} = 830 \times 1 \cdot 25 = 1038$ kN

Assume that each strand is stressed to 0·70 of its characteristic strength of 165 kN to give a prestressing force of $165 \times 0 \cdot 70 = 115 \cdot 5$ kN.

$$\text{No. of strands required} = \frac{1038}{115 \cdot 5} = 9$$

Nine strands of 12·5 mm nominal diameter will be provided. Let the area of strands be A_{ps}.

$$A_{ps} = 9 \times 94 \cdot 2 = 848 \text{ mm}^2$$

The ultimate limit state
Ultimate bending moment $= (1 \cdot 4 \times 350 + 1 \cdot 60 \times 450)$
$= 1200$ kN m.

Transition from fully prestressed to reinforced concrete 127

Let the area of the untensioned steel be $A_{s(u)}$. A trial neutral-axis depth of 12 cm from top will be assumed. Strain in prestressing tendon = strain due to effective prestress + strain in the concrete at tendon level due to prestress only + strain in tendon in the cracked section

$$= \frac{115 \cdot 5 \times 0 \cdot 80}{94 \cdot 2 \times 200} + \frac{115 \cdot 5 \times 9}{34 \times 100} \left(\frac{1}{3000} + \frac{31}{33\,500} \right)$$

$$+ \frac{0 \cdot 0035 \times 46 \cdot 6}{12 \cdot 0} = 0 \cdot 0049 + 0 \cdot 000\,68 + 0 \cdot 0136 = 0 \cdot 0192$$

where $E_s = 200 \text{ kN/mm}^2$; $E_c = 34 \text{ kN/mm}^2$ and the strain in extreme compression fibre is taken as $0 \cdot 0035$.

From Fig. 4.7 (reproduced from CP 110:1972), the strain in prestressing steel at $\frac{f_{pu}}{\gamma_m}$ is $= 0 \cdot 005 + \frac{165}{94 \cdot 2} \times \frac{1}{1 \cdot 15} \times \frac{1}{200}$

$$= 0 \cdot 0126.$$

The strain in the tendon of $0 \cdot 0192 > 0 \cdot 0126$. Obviously, the stress in the prestressing steel has reached (f_{pu}/γ_m). Hence the tensile force in 9 strands $= 9 \times \dfrac{165}{1 \cdot 15} = 1292 \text{ kN}$.

$$\text{Total compression} = 100 \times 12 \times 0 \cdot 4 \, f_{cu} = \frac{100 \times 12 \times 20 \times 100}{1000}$$

$$= 2400 \text{ kN}.$$

$$\left.\begin{array}{l}\text{Hence the balance to be made up}\\ \text{by untensioned steel}\end{array}\right\} = 2400 - 1292 = 1108 \text{ kN}$$

Let the untensioned steel be provided at an average lever arm of 64 cm from top.

$$\text{Strain in the untensioned steel} = \frac{0 \cdot 0035}{12} \times 64 = 0 \cdot 0187$$

Strain at f_y/γ_m from the stress-strain diagram given in Fig. 3 of CP 110 reproduced in this book as Fig. 4.8

$$= 0 \cdot 002 + \frac{425}{1 \cdot 15 \times 200 \times 1000} = 0 \cdot 003\,84.$$

Clearly, the steel would have reached a stress of f_y/γ_m.

$$A_{s(u)} = \frac{1108 \times 1000 \times 1 \cdot 15}{425 \times 100} = 30 \text{ cm}^2 = 3000 \text{ mm}^2$$

The tensioned and untensioned steels are arranged as shown in Fig. 8.4.

For computing the ultimate moment of resistance, it is convenient to replace the untensioned steel by an equivalent area of prestressing steel and find the centre of gravity of the total equivalent steel.

$$\left.\begin{array}{l}\text{Prestressing steel}\\ \text{equivalent of } A_{s(u)}\end{array}\right\} = 3000 \times \left(\frac{425}{1750}\right) = 730 \text{ mm}^2$$

$$\left.\begin{array}{l}\text{Depth of equivalent}\\ \text{c.g. from top}\end{array}\right\} = \frac{730 \times 64 + 848 \times 58 \cdot 6}{1578} = 61 \text{ cm}$$

$$\text{Lever arm} = 61 - \frac{12}{2} = 55 \text{ cm}$$

Hence $M_u = \frac{175}{1 \cdot 15} \times \frac{1578}{100} \times \frac{55}{100} = 1320$ kN m against the ultimate bending moment of 1200 kN m. This is quite satisfactory.

We may now check and see if the depth of neutral axis initially assumed is all right.

$$\frac{f_{pu}A_{ps}}{f_{cu}bd} = \frac{1750}{50} \times \left(\frac{15 \cdot 78}{60 \times 100}\right) = 0 \cdot 09$$

Interpolating from Table 37 of the Code, the ratio of $x/d = 0 \cdot 19$ or, the neutral axis depth $= 0 \cdot 19 \times 61 = 11 \cdot 6$ cm. This agrees closely with 12 cm assumed initially.

Serviceability limit state of cracking

$M_T = 800$ kN m; $\quad a_1 = 0 \cdot 19$; $\quad a_2 = 0 \cdot 30$; $\quad d_u = 64$ and $m = 6$

$$k_p = \left(\frac{80\,000 - 830 \times 31}{830} - 27 \cdot 6\right)\frac{1}{64} = 0 \cdot 59$$

$$k_1 = 3a_1(a_1 + 2k_p) = 3 \times 0 \cdot 19(0 \cdot 19 + 1 \cdot 18) = 0 \cdot 78$$

$$p_u = \frac{A_{s(u)}}{bd_u} = \frac{30 \cdot 0}{100 \times 64} = 0 \cdot 0047$$

$$K_2 = 6mp_u(1 + k_p) = 6 \times 6 \times 0 \cdot 0047(1 + 0 \cdot 59) = 0 \cdot 27$$

$$K_3 = a_1^2(2a_1 + 3k_p) = 0 \cdot 19^2(0 \cdot 38 + 1 \cdot 77) = 0 \cdot 0775$$

The cubic equation in n is

$$n^3 + 1 \cdot 77n^2 + 2 \cdot 72n - 1 \cdot 081 = 0$$

Solving, $n = 0 \cdot 32$

f_c is now calculated from (8.14)

$$f_c = \frac{830\,000}{100 \times 64} \left[\frac{2 \times 0.32}{0.19 \times (0.64 - 0.19) + 0.30 \times 0.13^2 - 2 \times 6 \times 0.0047 \times 0.68} \right]$$

$$= 1590 \text{ N/cm}^2$$

$$f_{s(u)} = \frac{6 \times 0.68 \times 1590}{0.32} = 20\,300 \text{ N/cm}^2$$

Using the CEB–FIP formula, the maximum crack-width for non-repetitive loads at the level of the untensioned steel is

$$(20\,300 - 4000) \times 10^{-6} = 0.0163 \text{ cm}$$

$$= 0.163 \text{ mm}$$

$$\left.\begin{array}{l}\text{Crack-width at soffit}\\ \text{level of the girder}\end{array}\right\} = \frac{0.163 \times (72 - 20.5)}{(64 - 20.5)}$$

$$= 0.193 \text{ mm} < 0.2 \text{ mm}$$

The crack-width computed by the more rigorous method given in Reference [8.10] is found to be 0.139 mm.

Example 8.2
Repeat the same problem with crack-width restricted to 0.10 mm and using the nominal tensile stress method given in CP 110:72.

It is clear from Example 8.1 that to restrict the crack-width to 0.10 mm either the prestress or the area of untensioned steel has to be increased. If the untensioned steel is increased, we may take advantage of the Code provision relating to the increase in the nominal tensile stress permitted for a crack-width of 0.10 mm. Of the two alternatives, increasing the prestressing force is likely to be the more effective. Let us, therefore, increase the number of strands from 9 to 14, keeping the area of untensioned steel unaltered.

Let us provide 14 strands to give an effective prestressing force of $14 \times 115 \cdot 5 \times 0 \cdot 80 = 1294$ kN.

$$\left.\begin{array}{l}\text{Nominal tensile stress}\\ \text{in bottom fibre}\end{array}\right\} = \frac{1\,294\,000}{3000}$$

$$+ \frac{1\,294\,000 \times 31}{33\,500} - \frac{80\,000\,000}{33\,500}$$

$$= 431 + 1195 - 2385 = 759 \text{ N/cm}^2$$

$$= 7.59 \text{ N/mm}^2$$

130 Modern prestressed concrete design

Restricting crack-width to 0·10 mm and taking into account the depth factor and the contribution of the untensioned steel, the allowable hypothetical tensile stress for the pretensioned member is

$$6 \cdot 3 \times 0 \cdot 84 + \frac{30}{3000} \times 100 \times 3 = 8 \cdot 30 \text{ N/mm}^2.$$

The design will just do.

8.7 Other Methods of Crack Control

Some codes of practice achieve control of crack-widths by either restricting the increase in the stress in the untensioned steel, the prestressed steel or both. Thus, for instance, in the Swiss Code SIA 162:1968, the following provisions are meant to control crack-widths. Increase in stress in prestressing steel is restricted to $0 \cdot 10 f_{pu}$ or 1500 kg/cm² for ordinary structures and $0 \cdot 05 p_u$ for bridges. Also, the stress in the reinforcing steel in the cracked section under working loads is restricted to 1500 kg/cm². Thürlimann considers the latter restriction generally conservative. The French Code ASP restricts the stress in the prestressing tendons under working loads to its initial value at transfer. In the ACI Code 318:1971, there is no specific mention of partial prestressing. But, where tensile stresses up to $12\sqrt{f'_c}$ are permitted, the Code demands a restriction on the deflexion which is computed by using the transformed section and bilinear moment-deflexion relationships. The restriction on deflexion indirectly serves to control crack-widths.

8.8 Deflexion of Class 3 Beams

Studies [8.9] have shown remarkable agreement between measured and computed values of the deflexion arrived at by using I_e the equivalent moment of inertia given in ACI 318:1971. The expression given in the ACI Code is

$$I_e = \left(\frac{M_{cr}}{M_{max}}\right)^3 I_g + \left[1 - \left(\frac{M_{cr}}{M_{max}}\right)^3\right] I_{cr}$$

where

M_{cr} = cracking moment

I_g = gross moment of inertia

M_{max} = maximum bending moment in the member

I_{cr} = moment of inertia of the cracked section.

We have already shown that $nd_u = d_n$. The depth of the neutral axis may be found from Equation (8.10). When once n is known

$$I_{cr} = \frac{1}{12} bd_n^3 + mA_{ps}(d_p - d_n)^2 + mA_{s(u)}(d_u - d_n)^2 \qquad (8.15)$$

8.9 Index for Degrees of Prestress

The degree of prestress in a beam may be expressed in one of two ways. It may be represented as the ratio of the tensile force contributed by the prestressed steel to the total tension at the ultimate limit state as suggested by Thürlimann [8.2]. The index i of the degree of prestress may therefore be written as

$$i = \frac{A_{ps}f_{p(0\cdot2)}}{A_{ps}f_{p(0\cdot2)} + A_{s(u)}f_{y(0\cdot2)}}, \qquad (8.16)$$

where $f_{p(0\cdot2)}$ and $f_{y(0\cdot2)}$ stand respectively for the 0·2% set proof stresses of the tensioned and untensioned steel. In the tests carried out by Thürlimann, the degree of prestress was characterized by this index.

An alternative means of describing the degree of prestress will be to compare the decompression moments, the beam with decompression occurring at the full working load being considered as fully prestressed. The beams in the experiments reported in Reference [8.9] were designated on this basis. Thus the degree of prestress in a beam in which decompression takes place at 60% of the working loads is 60%. In Fig. 8.2 which summarizes the load-deflexion behaviour of the test beams reported in [8.9], the designations GP 100, GP 80, etc., designate the degree of prestress.

References

[8.1] Freyssinet, E., 'Birth of prestressing', Speech on the occasion of his jubilee as an engineer, translated by A. J. Harris from French original in *Travaux*, July–August, 1954, Cement and Concrete Association, London.

[8.2] Thürlimann, B., and Calfisch, R., 'Teilweise vorgespannter Beton', Deutscher Beton Tag, 1969.

[8.3] Schweizerischer Ingenieur- und Architekten-Verein, *Norm für die Berechnung von Konstruktionen aus Beton, Stahlbeton und Spannbeton*, SIA Technische Norm 162, Zurich, 1968.

[8.4] Ramaswamy, G. S., and Raman, N. V., 'Optimum design of prestressed concrete sections for minimum cost by non-linear programming', Paper presented at the Sixth FIP Congress, Prague, June 1970.

[8.5] Abeles, P. W., and Czuprynski, 'Partial prestressing', *Annales des Travaux Publics de Belgique*, No. 2, 1966.

[8.6] Leonhardt, F., *Prestressed Concrete Design and Construction*, Wilhelm Ernst und Sohn, 2nd edn., translated into English by Amerongen, 1964.

[8.7] Bennett, E. W., and Chandrasekhar, C. S., 'Calculation of width of cracks in Class 3 prestressed concrete beams', *Proceedings of the Institution of Civil Engineers*, London, July 1971.

[8.8] Walther, R., and Bhal, N. S., 'Partial prestressing' (Prestressed Reinforced Concrete), University of Stuttgart, 1971. (In German; translation into English by Bhal, N. S.)

[8.9] Prasada Rao, A. S., Gandotra, K., and Ramaswamy, G. S., 'Flexural behaviour of beams prestressed to different degrees (Class 1 to Class 4 of CEB–FIP classification)', under publication.

[8.10] Parameswaran, V. S., Annamalai, G., and Ramaswamy, G. S., 'Theoretical and experimental investigations on the flexural behaviour of Class 3 beams', paper presented at the Seventh FIP Congress, New York, May 1974.

[8.11] Guyon, Y., *Limit State Design of Prestressed Concrete*, Vol. I, Translated into English by Chambon and F. H. Turner.

9
Statically indeterminate structures

In a statically determinate prestressed concrete structure the act of prestressing does not induce secondary or parasitic reactions. But in a statically indeterminate structure, external reactions are induced by prestressing, as is evident from the simple example of a two-span continuous beam prestressed by a single straight prestressing tendon (Fig. 9.1(a)). In Fig. 9.1(b), the parasitic bending moments are shown superimposed over the bending moments caused by prestressing to give the total bending moments due to prestress.

9.1 Concept of the Pressure Line

The concept of the pressure line or a C line has already been explained in Chapter 4. It may be recalled that in a statically determinate beam *carrying no load*, the pressure line coincides with the tendon itself. But not so for a statically indeterminate structure, where the pressure line shifts on account of the parasitic bending moments caused by prestressing. Reference to Examples 4.1 and 4.2 will clearly bring out this distinction between the behaviour of determinate and indeterminate systems.

9.2 Some Definitions and Theorems

The c.g.s. line (i.e., the tendon profile) of a continuous prestressed-concrete beam is said to be *linearly transformed*, if the tendon is moved to new positions over the interior supports, keeping the configuration of the cables (bends and curves) unaltered and keeping the eccentricities over the supports unchanged. For example, the pressure line of a continuous beam resulting from prestressing, shown in Fig. 9.1(a), represents a linear transformation of the tendon.

A tendon profile is said to be *concordant* if it gives rise to a coincident pressure or C line. It may be verified by means of examples, that if the prestressing tendon is aligned along a concordant profile, there will be no parasitic reactions or parasitic bending moments

134 Modern prestressed concrete design

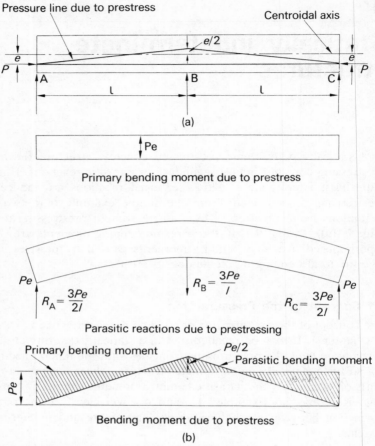

Fig. 9.1 Prestressing forces and moments

caused by prestressing. Any bending-moment diagram for a given beam due to any condition of loading on it will give rise to the shape of a concordant profile. It is not always necessary to select a concordant profile. Non-concordant profiles may also be equally satisfactory.

An important theorem may now be stated. In a prestressed continuous beam, any c.g.s. line can be linearly transformed without altering the pressure or C line. It follows that a linear transformation does not affect the stresses in the concrete.

9.3 Concept of Equivalent Loads

The effects of a prestressing tendon on a beam can be replaced by those of equivalent loads. A few simple examples will clarify this concept. In Fig. 9.2, a beam with a parabolic cable is shown. The eccentricity is e at midspan and zero at the ends. Hence the bending moment due to prestress is parabolic, being maximum at midspan and zero at the ends. It is easily verified that the bending moment M at a distance x from the left end may be written as

$$M = -\frac{4Pe}{l^2}x(1-x) \tag{9.1}$$

The *equivalent load* causing this bending moment can be found by differentiating M twice with respect to x to give

$$\frac{d^2M}{dx^2} = \frac{8Pe}{l^2} \tag{9.2}$$

This equivalent load can be used instead of the prestressing force to find the shear force, bending moment, deflexion, etc., caused by the prestressing force.

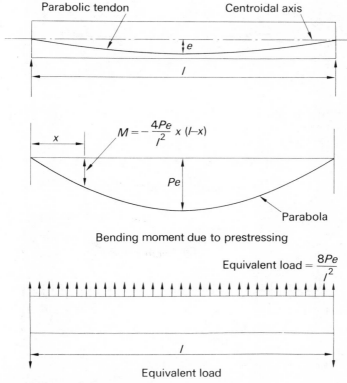

Fig. 9.2 Beam with parabolic tendon

136 Modern prestressed concrete design

If a straight prestressing tendon is used, the equivalent loads are end couples of magnitude Pe.

The next example considered (Fig. 9.3) is that of a beam in which the eccentricities are e_1 and e_2 at the ends and e_3 at midspan. Taking co-ordinates as shown, the equation of the cable profile which is parabolic may be expressed as

$$e = \left(\frac{2e_2 - 4e_3 + 2e_1}{l^2}\right)x^2$$
$$+ \frac{(4e_3 - e_2 - 3e_1)}{l}x + e_1 \qquad (9.3)$$

The bending moment M at a point distant x from the left end is obviously Pe and hence

$$M = -Pe \qquad (9.4)$$

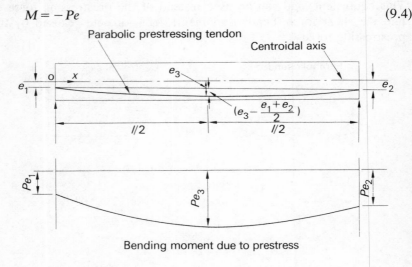

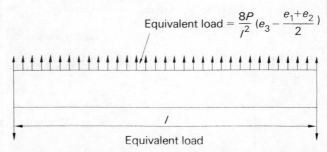

Fig. 9.3 Beam with parabolic tendon and with eccentricities

Statically indeterminate structures   137

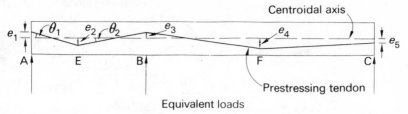
Equivalent loads

Fig. 9.4 Cable trajectory of a cantilever beam with sharp changes in direction

The equivalent load q is given by d^2M/dx^2. Differentiating Equation (9.3) and multiplying by P, we get

$$q = -\frac{(4e_2 - 8e_3 + 4e_1)}{l^2} P \qquad (9.5)$$

$$= \frac{8P}{l^2}\left(e_3 - \frac{e_1 + e_2}{2}\right) \qquad (9.6)$$

It is easily verified that $\left(e_3 - \dfrac{e_1+e_2}{2}\right)$ *is the vertical intercept at midspan between the line joining the end points of the cable and the tendon trajectory.* Equations (9.2) and (9.6) for equivalent loads are of the same form if the vertical intercept mentioned above is inserted in (9.6) to take the place of e in (9.2).

In Fig. 9.4, the cable trajectory of a continuous beam with sharp changes in direction is shown. The equivalent load at $E = P(\sin \theta_1 + \sin \theta_2)$. Noting that the cable profile is flat, $\sin \theta_1 \approx \tan \theta_1$; $\sin \theta_2 \approx \tan \theta_2$. Hence, the equivalent load at E is obviously

$$P(\tan \theta_1 + \tan \theta_2)\uparrow = P\left(\frac{e_1+e_2}{AE} + \frac{e_2+e_3}{BE}\right) \qquad (9.7)$$

Similarly, the equivalent load at

$$F = P\left(\frac{e_3+e_4}{BF} + \frac{e_4-e_5}{CF}\right)\uparrow \qquad (9.8)$$

Design Example 9.1
A prestressed-concrete beam continuous over two spans carries a uniformly-distributed load of 20 kN/m (Fig. 9.5). Locate the C line due to prestressing and external loading. Prestressing force = 1500 kN.

Solution
The primary bending-moment diagram due to prestressing is drawn by computing Pe at various sections (Fig. 9.5(a)). The loads resulting from

138 Modern prestressed concrete design

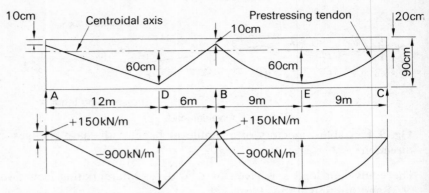

Fig. 9.5(a) Primary bending-moment diagram due to prestress at various sections

prestressing may be found by simple statics. For example, bending moment at A due to prestressing is +150 kN m (sagging). At D the bending moment is −900 kN m (hogging). The change is (900+150) = 1050. Hence a force must act down at A to cause a hogging moment of 1050 kN m at D at a distance of 12 metres. Obviously, the force at A = 1050/12 = 87·5 kN↓ . Similar reasoning will show that a load of 262·5 kN↑ has to act at D. Now starting from C, it is to be noted that the shape of the bending-moment diagram is parabolic indicating that a uniformly distributed load has to act on span BC. The intensity of the load is found by using equation (9.6) as

$$\frac{8 \times \left[60 - \left(\frac{-10+0}{2}\right)\right] \times 1500}{18 \times 18 \times 100} = 24 \text{ kN/m} \uparrow$$

Next, the reaction at C (assuming it to be downward) is found by setting up the expression for the bending moment at E as

$$-9R_C + \frac{24 \times 9 \times 9}{2} = -900$$

$$R_C = 208 \text{ kN} \downarrow$$

Fig. 9.5(b) Equivalent loads acting on the beam on account of prestress

Statically indeterminate structures 139

Now equating the sum of all vertical forces caused by prestressing acting on the beam to zero, $87 \cdot 5 - (24 \times 18) - 262 \cdot 5 + 208 + R_B = 0$, $R_B = 399$ kN↓. The forces acting on the beam on account of prestressing have now been completely determined and they are shown in Fig. 9.5(b).

Alternatively, the forces acting on the beam could have been found as follows:

$$\left.\begin{array}{l}\text{Reaction } R_A \text{ at A by}\\\text{using equation (9.8)}\end{array}\right\} = 1500 \times \frac{(60+10)}{1200}$$

$$= 87 \cdot 5 \text{ kN}\uparrow$$

$$\left.\begin{array}{l}\text{Equivalent load at D}\\\text{by equation (9.8)}\end{array}\right\} = 1500\left(\frac{70}{1200} + \frac{70}{600}\right)$$

$$= \frac{1500}{1200} \times 210 = 262 \cdot 5 \text{ kN}\uparrow$$

$$\left.\begin{array}{l}\text{Equivalent load at B}\\\text{due to force from}\\\text{span AB by equation}\\\text{(9.8)}\end{array}\right\} = \frac{1500 \times 70}{600} = 175 \text{ kN}\downarrow$$

In addition, there will be a force transferred from span BC at B which may be computed as $1500 \times$ slope of tendon in span BC at B. The slope of the tendon may be found by setting up the equation of the tendon profile and finding its derivative at B. From equation (9.3), the equation of the parabolic profile in span BC may be found as

$$e = \left(\frac{240 + 20}{1800^2}\right)x^2 + \left(\frac{-240 - 30}{1800}\right)x + 10$$

$$\frac{de}{dx} = \frac{520}{1800^2}x - \frac{270}{1800}$$

$\frac{de}{dx}$ at B, i.e., at $x = 0$ is $-\frac{270}{1800}$.

$$\left.\begin{array}{l}\text{Hence the reaction at B due}\\\text{to tendon in span B}\end{array}\right. = 1500 \times \frac{270}{1800} = 225 \text{ kN}\downarrow$$

Hence total force at B $= 225 + 175 = 400$ kN↓

The reaction at C $= 1500 \times \left(\frac{de}{dx}\right)_{x=1800}$

$$= 1500 \times \frac{250}{1800} = 208 \cdot 3 \text{ kN}\downarrow$$

These reactions agree with the forces already found by simple statics.

The secondary bending moments caused by prestressing are easily found by moment distribution as follows.

A		B		C
+150	−350	+700	−648	+648
0	+200	0	0	−648
+150	−150	+100	−324	0
		+800	−972	
		+86	+86	
		+886	−886	

From the results of the moment distribution, it is inferred that a sagging moment of 886 kN m will be caused at the support B. Hence the pressure line at B will move $(886-150)/1500 = 0{\cdot}497$ m $= 49{\cdot}7$ cm above the tendon at that point, due to the effect of prestressing only. The bending-moment diagram due to prestressing only may now be drawn (Fig. 9.5(c)). From the bending-moment diagram, the shift in the pressure line on account of prestressing can be found. For example, at a section 12 m from A, the bending moment due to prestressing from Fig. 9.5(c) is found to be 409 kN m (hogging). Hence the eccentricity of the pressure line at this section is $(409 \times 100)/1500 = 27{\cdot}3$ cm below the centroid of the section. But the tendon is at 60 cm below the centroid. Hence the pressure line at this section will be $(60-27{\cdot}3) = 32{\cdot}7$ cm above the c.g.s. of the tendon. Similarly, the position of the pressure line due to prestressing at other sections may be established.

Shift in pressure line due to loads

So far, we have considered only the bending moments resulting from prestressing. The bending moments resulting from the loads may be

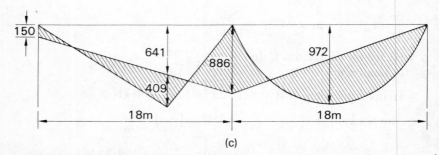

Fig. 9.5(c) Moments due to prestressing in kN m

found by a simple moment-distribution as shown below

A	B		C
+540	−540	+540	−540
−540	0	0	+540
	−270	+270	
0	−810	+810	0

The bending moment due to loading is shown in Fig. 9.5(d). It will be seen that loading causes a hogging bending moment of 810 kN m over the support B. We have already seen that prestressing gave rise to a sagging moment of 886 kN m. Superimposing both the effects, the net bending moment at B is a sagging moment of 76 kN m. The eccentricity of the pressure line at B due to *prestressing and loading* is therefore $(76/1500) \times 100 = 5 \cdot 67$ cm *above* the centroid of the section. The location of the pressure line at other sections may be found similarly. This is left as an exercise for the reader.

9.4 Load-balancing

In 1961, T. Y. Lin [9.1] proposed a new method for the design of prestressed beams. It offers a simple and elegant means for the design of statically indeterminate systems such as continuous beams, rigid frames, slabs and shells. It does not offer any significant advantages when applied to simply-supported beams. However, the simply-supported beam will be used to explain the underlying principle.

Consider the simply-supported beam in Fig. 9.2. If the prestressing force is P, we have already shown that the tendon exerts uniformly-distributed upward load of $8Pe/l^2$ on the beam. Let us suppose that the beam carries a uniformly-distributed load of q. This downward load of

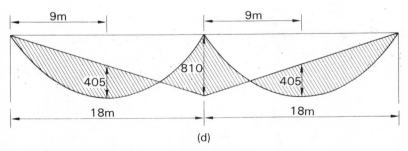

(d)

Fig. 9.5(d) Bending moment due to loads in kN m

142 Modern prestressed concrete design

q can be completely balanced by the upward load caused by prestressing, by choosing P such that $8Pe/l^2 = q$ or,

$$P = \left(\frac{ql^2}{8e}\right) \tag{9.9}$$

If the cable has different eccentricities at its ends, instead of e, the intercept between the tendon and the straight line joining the ends of the tendon has to be used as already explained. If load on the beam is thus balanced, the beam is not subjected to any bending moment and therefore flexural stresses in the beam will be zero. There will be only a uniform axial compressive stress caused by the prestressing force. It is quite clear that in such a situation, the pressure line has to coincide with the centroidal axis. The deflexions of the beam are zero. Concentrated loads acting on a beam can also be balanced and the appropriate prestressing force needed can be found as follows.

Referring to Fig. 9.6, the force exerted by the tendon at $C = Pe\left(\frac{1}{a} + \frac{1}{b}\right)$. If this force is equated to the load at C,

$$W = \frac{Pe(a+b)}{ab} = \frac{Pel}{ab}$$

or

$$P = \left(\frac{Wab}{el}\right) \tag{9.10}$$

Another case of practical importance is a cantilever with a prestressing tendon in the form of a parabola having no eccentricity and a horizontal tangent at the free end and an eccentricity of e at the fixed end. Let the cantilever carry a uniformly distributed load q (Fig. 9.7). The problem is to find the appropriate prestressing force to balance it. It is easily verified that the equation of the parabola is $y = (e/l^2)x^2$, x being

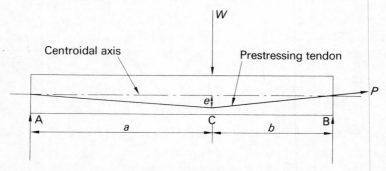

Fig. 9.6 Load-balancing of concentrated loads

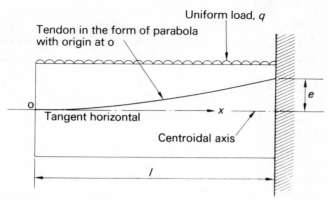

Fig. 9.7 Balancing uniform load on a cantilever

measured from origin O. The bending moment at a section distant x from O is

$$M = \frac{Pe}{l^2} x^2,$$

P being the prestressing force in the tendon. Two successive differentiations give the upward load caused by prestressing as $2Pe/l^2$. If the load on the beam is to be balanced,

$$\frac{2Pe}{l^2} = q$$

or

$$P = \frac{ql^2}{2e} \tag{9.11}$$

The next question to be answered is, 'What is the part of the total load to be balanced by prestressing?'. In most structures, the full live loads prescribed in the codes occur only infrequently. Hence, if the load to be balanced is selected as the dead load plus the full live load, the structure will have a permanent upward camber and this camber will further increase in course of time on account of creep. The appropriate load to be balanced is, therefore, the dead load and the fraction of the live load that occurs frequently. The load-balancing principle can be understood easily when it is applied to an actual problem.

Design Example 9.2
The continuous beam shown in Fig. 9.8(a) carries a uniformly distributed load of 20 kN/m. Find the prestressing force and an appropriate trajectory for the tendon.

144 Modern prestressed concrete design

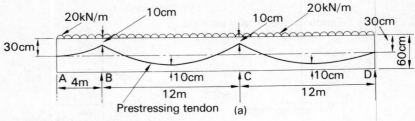

Fig. 9.8(a) Continuous beam

Solution
Loads being uniformly distributed, parabolic tendons as sketched in (Fig. 9.8(a)) would be suitable.

Cantilever AB

$$\left.\begin{array}{l}\text{Prestressing force } P\\ \text{required using}\\ \text{equation (9.11)}\end{array}\right\} = \frac{20 \times 4 \times 4}{0.2 \times 2} = 800 \text{ kN}$$

Span BC

$$\left.\begin{array}{l}\text{Prestressing force}\\ \text{required using}\\ \text{equation (9.9)}\end{array}\right\} = \frac{20 \times 12 \times 12}{8 \times 0.40} = 900 \text{ kN}$$

Span CD

$$\left.\begin{array}{l}\text{Prestressing force}\\ \text{required by using}\\ \text{equation (9.9) and}\\ \text{using the intercept}\\ \text{between the tendon}\\ \text{and the line joining}\\ \text{its ends for } e\end{array}\right\} = \frac{20 \times 12 \times 12}{8 \times 0.30} = 1200 \text{ kN}$$

If the cable is stretched from the end D, a prestressing force of 1200 kN at D will be appropriate for all spans on account of friction The prestressing force in the other spans will be less. The frictio losses are computed as follows.

In Fig. 9.8(b), the slopes of the cable in radians are shown. T cumulative slope is

$$\alpha = 0.23 + 0.26 + 0.26 + 0.20 = 0.95 \text{ radians}$$

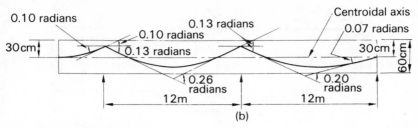

Fig. 9.8(b) Slopes of tangents

Total length of tendon $l = 28$ m

$\mu\alpha = 0.30 \times 0.95 = 0.285$

$kl = 20 \times 10^{-4} \times 28 = 0.056$

$\mu\alpha + kl = 0.341$

Let prestressing force at A be P_x and the prestressing force at D = P_0. $P_x = P_0 e^{-0.341}$. Or, $1.40\ P_x = P_0$.

$P_0 = 1200$ kN

$P_x = \dfrac{1200}{1.40} = 858$ kN.

Hence continuing the same tendon in all the spans is quite satisfactory.

Design Example 9.3
Design a prestressed continuous beam of two equal spans of 18 m, effective span, framing into columns 60 cm × 60 cm and 4 m long. The characteristic strength of the concrete is prescribed as 50 N/mm². Beams are spaced 4 m apart (Fig. 9.9(a)).

Solution
The preliminary cross-section of the beam is assumed as shown in Fig. 9.9(b).

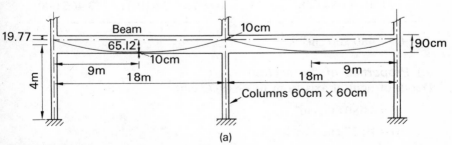

Fig. 9.9(a) Continuous T beam

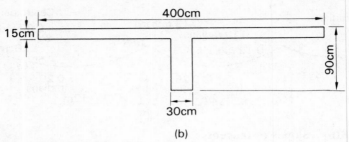

Fig. 9.9(b) Cross-section

(1) Effective width of flange
The effective width may be assumed as the lesser of the two values
 (a) $\frac{1}{4}$ span $= \frac{1}{4} \times 1800 = 450$ cm
 (b) Spacing centre to centre $= 400$ cm

(2) Neutral axis position

No	A	x	Ax
1	6000	7·5	45 000
2	2250	52·5	118 125

$\sum A = 8250 \quad \sum Ax = 163\,125$

$\dfrac{\sum Ax}{\sum A} = \dfrac{163\,125}{8250} = 19 \cdot 77$ cm from top.

No.	A	$bd^3/12$	s	s^2	As^2	I(Total), cm^4
1	6000	112 500	12·27	150·55	903 300	1 015 800
2	2250	1 054 688	32·73	1071·25	2 410 312	3 465 000

$I = 4\,480\,800$ cm^4

(4) Properties of the section
Depth of neutral axis from top $= 19 \cdot 77$ cm
   $I = 4\,480\,800$ cm^4
   $y_t = 19 \cdot 77$ cm
   $y_b = 70 \cdot 23$ cm

Statically indeterminate structures 147

$Z_t = 226\,646 \text{ cm}^3$

$Z_b = 63\,802 \text{ cm}^3$

(5) Bending moments

Self-weight/metre $= \dfrac{0{\cdot}8250 \times 23\,000}{1000} = 19 \text{ kN/m}$

Assume a load of 250 N/m² to account for finishes and waterproofing.

$\left.\begin{array}{l}\text{Load/m due to finishes}\\ \text{and waterproofing}\end{array}\right\} = \dfrac{250 \times 4}{1000} = 1 \text{ kN/m}$

Total dead load $\qquad = 20 \text{ kN/m}$

Live load $\qquad\qquad = 2{\cdot}5 \text{ kN/m}^2 = 10 \text{ kN/m}$

As a preliminary to frame analysis by moment distribution, the I/l for the beam and the column are computed as follows.

$\dfrac{I}{l}$ for beam $= \dfrac{4\,480\,800}{1800} = 2490 \text{ cm}^3$

I of column $= \dfrac{1}{12} \times 60 \times 60^3 = 1\,080\,000 \text{ cm}^4$

$\dfrac{I}{l}$ for column $= \dfrac{1\,080\,000}{400} = 2700 \text{ cm}^3$

$\dfrac{I}{l}$ for column: $\dfrac{I}{l}$ of beam $\approx 9:8$.

The bending moments due to dead load are computed as follows by moment distribution (Fig. 9.9(c)):

		−667	+667		
	+286	−127	+127	−286	
	−254	0	0	+254	
	+540	−540	+540	−540	

(left column: −286, 0, −286, 0 / −143, −143, 0, 0)

(right column: 0, +286, +286, 0 / +143, +143, 0, 0)

(Sign convention: clockwise over joint positive)

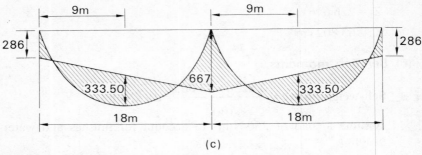

Fig. 9.9(c) Bending moment due to loads in kN m

Similarly, the bending moments due to live load are also found by moment distribution. Because the live load is half the dead load, the live-load moments will be just one half of the bending moments due to dead load. Hence the total bending moments due to dead and live loads are simply found by multiplying the dead load bending moments by 1·5 to get the following values:

+429	−1000·5	+1000·5	−429
−214·5 −429			+214·5 +429

(6) Prestressing force

The prestressing force will be found by balancing the dead load and half the live load, i.e., 25 kN/m. The intercept at midspan between the line joining the ends of the tendon and the tendon = $(70\cdot23 + 80)/2 - 10 = 65\cdot12$ cm. Hence the prestressing force P required $= \dfrac{25 \times 18 \times 18 \times 100}{8 \times 65\cdot12}$

$= 1554$ kN

Axial compression due to prestress $\} = \dfrac{1554}{8250} \times \dfrac{1000}{100} = 1\cdot89 \text{ N/mm}^2$

This is satisfactory.

In this problem, it would appear to be enough if the prestress balances only the dead load because the sectional properties of the beam are quite ample and the stresses are likely to be within limits.

$$\left.\begin{array}{l}\text{Prestressing force } P\\ \text{required to balance}\\ \text{dead load only}\end{array}\right\} = \frac{20 \times 18 \times 18 \times 100}{8 \times 65 \cdot 12}$$

$$= 1243 \cdot 2 \text{ kN}$$

(7) Analysis of sections

(a) Exterior support section

$$\left.\begin{array}{l}\text{Residual bending moment}\\ \text{at exterior support}\\ \text{after load-balancing}\\ \text{i.e., the live-load moment}\end{array}\right\} = 143 \text{ kN m}$$

Tensile stress at top $= \dfrac{143\ 000}{226\ 646} = 0 \cdot 631 \text{ N/mm}^2$

$$\left.\begin{array}{l}\text{Axial compression due to}\\ \text{prestress}\end{array}\right\} = \frac{1243 \cdot 2}{8250} \times \frac{1000}{100}$$

$$= 1 \cdot 5 \text{ N/mm}^2$$

Superimposing the two, there is a compressive stress of $0 \cdot 87 \text{ N/mm}^2$ at the top fibre at the exterior support.

(b) Midspan section

Residual B.M. $= 166 \cdot 75 \text{ kN m}$

Tensile stress at bottom $= \dfrac{166 \cdot 75 \times 1000}{63\ 802}$

$$= 2 \cdot 61 \text{ N/mm}^2$$

$$\left.\begin{array}{l}\text{Axial compression due}\\ \text{to prestress}\end{array}\right\} = 1 \cdot 50 \text{ N/mm}^2$$

Net tension $= 1 \cdot 11 \text{ N/mm}^2$

This is satisfactory.

(c) Interior support section

Residual bending moment $= 333 \cdot 5 \text{ kN m}$

Tensile stress at top $= \dfrac{333 \cdot 5 \times 1000}{226\ 646}$

$$= 1 \cdot 47 \text{ N/mm}^2$$

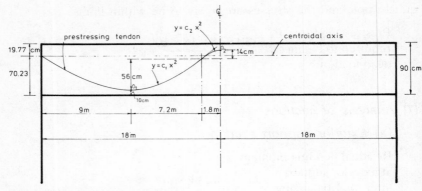

Fig. 9.10 Final cable profile

$$\left.\begin{array}{l}\text{Axial compression due}\\\text{to prestress}\end{array}\right\} = 1\cdot 50 \text{ N/mm}^2$$

Hence net stress on the top fibre is zero.
Hence the prestress provided is adequate at all sections.

(8) *Final cable profile*

To blunt the sharp bends of the tendon over the interior supports, the modified cable profile shown in Fig. 9.10 consisting of two parabolas $y = c_1 x^2$ and $y = c_2 x^2$ with their origins at O_1 and O_2 may be adopted. The two curves are such that their slopes are the same at the point of inflexion which is 1·8 m from the interior support. The chosen tendon trajectory will result in upward and downward forces as shown in Fig. 9.11. The upward and downward forces are computed as follows

$$\text{upward force} = \frac{8 \times 1243\cdot 2 \times 56}{100 \times 16\cdot 2 \times 16\cdot 2}$$

$$= 21\cdot 22 \text{ kN/m}$$

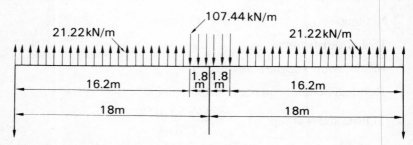

Fig. 9.11 Forces exerted by cable

$$\text{downward force} = \frac{8 \times 1243 \cdot 2 \times 14}{100 \times 3 \cdot 6 \times 3 \cdot 6}$$

$$= 107 \cdot 44 \text{ kN/m}$$

The moments caused by prestressing may now be computed by moment distribution. Using the formula for the fixed-end moment for partially loaded beams (Fig. 9.12), the f.e.ms are computed for the upward and downward loads and moment distribution is carried out as shown below. The f.e.ms at the exterior support and interior support of the left-hand span are $+560$ kN m and $-391 \cdot 26$ kN m respectively.

(Sign convention: Clockwise over joint positive)

+296·48	−523·02	+523·02	−296·48
0	−131·76	+131·76	0
−263·52	0	0	+263·52
+560·00	−391·26	+391·26	−560·00

A   B   C

For purposes of finding the reaction R_A at A, these bending moments will have to be designated as positive when they cause sagging, and negative when they result in hogging. Making this change, assuming R_A acts downward

$$-18R_A + 296 \cdot 48 + (21 \cdot 22 \times 16 \cdot 2 \times 9 \cdot 9)$$

$$-(107 \cdot 44 \times 1 \cdot 8 \times 0 \cdot 90) = +523 \cdot 02$$

$$R_A = 166 \cdot 83 \text{ kN}$$

B.M. at midspan due to prestressing $\Big\} = -\tfrac{18}{2} \times 166 \cdot 83$

$+296 \cdot 48 + (21 \cdot 22 \times 9 \times 4 \cdot 5) = -346 \cdot 5$ kN m

We are now ready to superimpose the bending moments due to prestressing and loading at the exterior support, midspan and interior

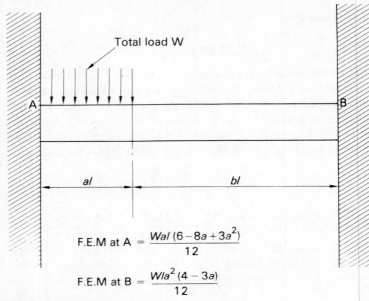

Fig. 9.12 Fixed-end moments

support sections and check these sections:

(a) *Exterior support section*

Bending moment due to prestressing $= +296\cdot 48$ kN m

Bending moment due to loading $= -429$ kN m

Hence net bending moment $= -132\cdot 52$ kN m

Tensile stress at top fibre due to bending $= \dfrac{132\cdot 52 \times 1000}{226\,646}$

$= 0\cdot 584$ N/mm^2

Axial compression due to prestress $= 1\cdot 50$ N/mm^2

Stress in top fibre $= 0\cdot 916$ N/mm^2 (compression)

This is acceptable.

Statically indeterminate structures

(b) Midspan section

B.M. due to prestressing = $-346{\cdot}5$ kN m

B.M. due to loading = $+500{\cdot}25$ kN m

Net bending moment = $+153{\cdot}75$ kN m

Tensile stress due to bending in the bottom fibre $= \dfrac{153{\cdot}75 \times 1000}{63\,802}$

$= 2{\cdot}4$ N/mm^2

Axial compression due to prestressing $= 1{\cdot}50$ N/mm^2

Hence tensile stress in bottom fibre $= 0{\cdot}90$ N/mm^2

This is within permissible limits.

(c) Interior support section

Bending moment due to prestressing $= +523{\cdot}02$ kN m

Bending moment due to loading $= -1000{\cdot}5$ kN m

Hence net bending moment $= -477{\cdot}48$ kN m

Tensile stress at the top fibre due to bending $= \dfrac{477{\cdot}48 \times 1000}{226\,646}$

$= 2{\cdot}11$ N/mm^2

Compression due to prestressing $= 1{\cdot}5$ N/mm^2

Hence net tensile stress at top fibre $= 0{\cdot}61$ N/mm^2

This is well within permissible limits. The chosen cable profile is satisfactory.

To obtain the desired prestressing force of $1243{\cdot}2$ kN at the interior support section, it is necessary to apply at the exterior support section a force of

$$1243{\cdot}2 \times 1{\cdot}1794 \times 1{\cdot}25 = 1833{\cdot}0 \text{ kN}$$

The first factor accounts for the friction and the second for other losses. Let us use 7 wire strands of nominal diameter $12{\cdot}5$ mm and

154 Modern prestressed concrete design

characteristic strength of 165 kN. Stressing the strand to 0·70 of its characteristic strength, the force provided by each strand = 115·5 kN.

$$\text{Number of strands required} = \frac{1833 \cdot 0}{115 \cdot 5} = 16.$$

9.5 Load Balancing Applied to Slabs

The concept of load-balancing may be applied, with advantage, to the design of slabs. The camber control of lift-slabs is achieved by balancing the dead load and that fraction of the prescribed live load which occurs frequently. By this means, it is possible to ensure that the slab remains level in service.

The principle of load-balancing developed for beams when applied for a slab takes the form

$$\frac{8P_1 e_1}{l_1^2} + \frac{8P_2 e_2}{l_2^2} = q \tag{9.12}$$

where q is the uniformly distributed load to be balanced; e_1 and e_2 are the sags of the cable in the two directions of span l_1 and l_2; P_1 and P_2 are the corrresponding prestressing forces. It is clear that a number of different combinations of P_1 and P_2 are possible. Generally, it is found economical to carry the bulk of the load in the shorter direction.

Design Example 9.4

Problem: A two-way slab $10 \text{ m} \times 15 \text{ m}$ simply-supported on walls is to carry a live load of 2 kN/m^2. Arrive at the prestressing forces and cable profiles.

Solution
Let it be assumed that the slab is 20 cm thick. Dead weight of slab $= \dfrac{0 \cdot 20 \times 23\,000}{1000} = 4 \cdot 6 \text{ kN/m}^2$. Let one fifth of the specified live load be balanced. Hence the total load to be balanced $= 4 \cdot 6 + \frac{2}{5} = 5 \text{ kN/m}^2$. As already noted, it is advantageous to carry the major part of the load in the shorter direction. Let us assume that it is enough to have a uniform compression of 2 N/mm^2 in the longer direction. Hence the prestressing force per metre length is

$$\frac{2 \times 10^6 \times 0 \cdot 2 \times 1}{1000} = 400 \text{ kN/m}$$

Let the sag of the cable be 6 cm (Fig. 9.13)
Hence,

$$400 = \frac{q_1 \times 15 \times 15}{8 \times 0 \cdot 6}$$

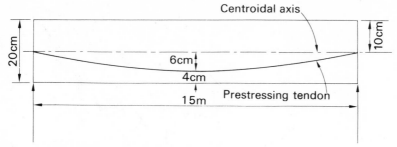

Fig. 9.13 Load balancing of slabs

or

$q_1 = 0.85$ N/mm^2.

The remaining load of $(5 - 0.85) = 4.15$ N/m^2 is to be carried in the shorter direction.

$$\left.\begin{array}{l}\text{Hence prestressing force}\\ \text{in the shorter direction}\end{array}\right\} = \frac{4.15 \times 10 \times 10}{8 \times 0.06}$$

$$= 864.58 \text{ kN}$$

$$\left.\begin{array}{l}\text{Compressive stress in}\\ \text{shorter direction}\end{array}\right\} = \frac{864.58 \times 1000}{20 \times 100 \times 100}$$

$$= 4.323 \text{ N/mm}^2$$

9.6 Load Balancing Applied to Shells

In long cylindrical shells, the tension in the longitudinal edge beams tends to be large, calling for the provision of heavy reinforcement. The placing and welding of such reinforcements poses difficult problems in construction. One answer is to prestress the edge beams. An even more satisfactory solution is to prestress the part of the valley of the shell near its junction with the edge beams (Fig. 9.14). A long cylindrical shell is known to behave more or less like a beam. If the weight of the shell per unit length along the span is q and h is the vertical projection of the sag of the tendon at midspan, the prestressing force required is

$$P = \frac{ql^2}{8h}$$

The advantages of prestressing the shell in the manner described are twofold. First, the deflexion of the shell is practically eliminated and it behaves practically as a pure membrane. Second, the shell will develop

156 Modern prestressed concrete design

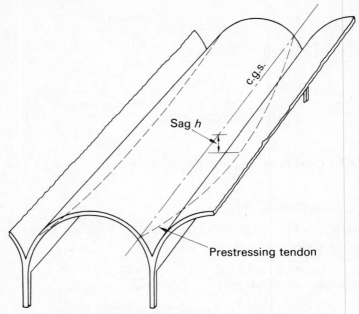

Fig. 9.14 Sag of prestressing cable in a cylindrical shell

a uniform state of longitudinal compression. Transverse bending is not completely eliminated. Traverse bending moments may be found by arch analysis. The load-balancing concept can also be applied to continuous and cantilever shells and folded plates [9.2].

References

[9.1] Lin, T. Y., 'A new concept for prestressed concrete', *Journal of the American Concrete Institute*, December 1961.

[9.2] Lin, T. Y., and Kulka, Felix, 'Concrete shells prestressed for load-balancing', *Proceedings, World Conference on Shell Structures*, October, 1962, San Francisco, pp. 423–30.

10
Optimum design

The advent of the electronic digital computer has revolutionized structural analysis. During the past two decades, powerful analytical tools have also become available to match the power of the computer. These two developments have made the optimum design of structures without undue expenditure of time, energy and money, a practical possibility.

In aerospace structures, the motivation for optimization stems from the need to *minimize* weight. Aeronautical engineers were, therefore, the first to think in terms of optimized designs leading to minimum weight. In civil engineering structures, weight is not an overriding consideration. Minimization of cost or consumption of materials are often the desired objectives.

10.1 Some Definitions

The function that is sought to be minimized is known as the *objective* or *merit* function. It offers a basis for weighing the relative merits of alternative designs. The objective function may be weight for aerospace structures and cost for civil engineering structures.

The search for an optimum design is, however, circumscribed by the need to ensure that the structure fully meets the functions for which it is intended, i.e., it has adequate strength, the stresses in it are within the values prescribed by specifications, its deflexions are within tolerable limits and its geometrical dimensions are aesthetically and functionally acceptable and so on. These restrictions are known as *constraints*. Those relating to structural behaviour, e.g., limits on stresses, deflexions, etc., are known as *behavioural constraints*. Those which specify minimum and maximum dimensions are often described as *side constraints*.

In problems of structural analysis, some of the parameters are already given or specified, e.g., the span, the loading and the strength properties of the material used. These are known as *preassigned parameters*.

158 Modern prestressed concrete design

These variables which are not preassigned and left free are known as *design variables*. They may be represented by the column vector

$$X = \begin{bmatrix} x_1 \\ x_2 \\ x_3 \\ \cdot \\ \cdot \\ \cdot \\ x_n \end{bmatrix}$$

Each design represented by X is a point in hyperspace defined by the design variables.

10.2 Statement of the Optimization Problem

The optimization problem may now be stated thus:

Minimize the objective function $F(X)$, under constraints
$G_j(X) \leq 0$, $j = 1, 2, \ldots M$.

If both the objective function as well as the constraints are linear, optimization leads to a *linear programming problem* which may be solved by using the Simplex algorithm. On the other hand, if the objective function or any of the constraints is non-linear, a *non-linear programming problem* will result.

10.3 Methods of Solution

In recent years, several powerful techniques have been developed to solve the non-linear programming problem involved in optimization [10.1]. Three of these techniques are mentioned below:

(1) Method of direct search
(2) Linearization of constraints and the objective function by Taylor series in the neighbourhood of a design point to reduce the non-linear programming problem to a linear programming problem, and
(3) The sequential unconstrained minimization technique (SUMT)

10.4 Method of Direct Search

Of these, the method of direct search is simple and straightforward. But its limitation is that there should be only two design variables defining a two-dimensional design space permitting the objective function and constraints to be graphically represented. Each candidate design is a point in the design space. This method permits the optimization process to be clearly visualized.

Linearization by the Taylor series method is also very useful even in solving complex optimization problems with numerous constraints [10.2].

In the SUMT procedure [10.3], the constrained minimization problem is replaced by a sequence of unconstrained minimization problems. The conversion is effected by introducing an interior or exterior *penalty function*. If an interior penalty function is used, successive design points considered, including the initial design point, must belong to the feasible region. If an exterior penalty function is used, the candidate design points must lie in the unfeasible region.

If, in the SUMT, an interior penalty function is employed, the optimization problem may be stated thus:
Minimize

$$\phi(X, R) = F(X) - R \sum_{j=1}^{M} \frac{1}{G_j(X)}$$

where $\phi(X, R)$ is the penalized objective function to be minimized and R is the penalty parameter. It is to be noted that the form of the function is such that it tends to be infinite or 'blows up', if any of the constraints is approached. The sequential unconstrained minimization of $\phi(X, R)$ for successively decreasing values of R forces $F(X)$ to its optimum or minimum value.

The direct search method and SUMT are best understood when applied to an actual problem stated below.

Design Example 10.1

(a) The direct search method

A pretensioned Class 1 beam of 12 metres span carries a superimposed load of $3000 \, \text{N/m}^2$. The characteristic cube strength of the concrete specified is $50 \, \text{N/mm}^2$. The prestressing 7 wire strands to be used are of 15·2 mm diameter (cross-sectional area = $138·7 \, \text{mm}^2$) and characteristic strength is 227 kN. The cost of concrete per cubic metre and prestressing steel per tonne may be taken as 700 p and 10 000 p respectively. The loss in prestress may be assumed as 22%. The preassigned parameters are shown in Fig. 10.1. The design is to be governed by CP 110:1972 and the maximum deflexion is limited to span/360.

Arrive at an optimum design using the method of direct search, given that the minimum depth of web $DW_{min} = 20 \, \text{cm}$ and the minimum prestressing force $PF_{min} = 400 \, \text{kN}$. The strands are initially stressed to 60% of their characteristic strength.

160 Modern prestressed concrete design

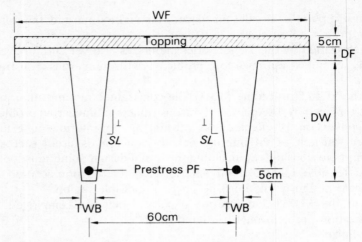

Fig. 10.1 Double T section with design variables

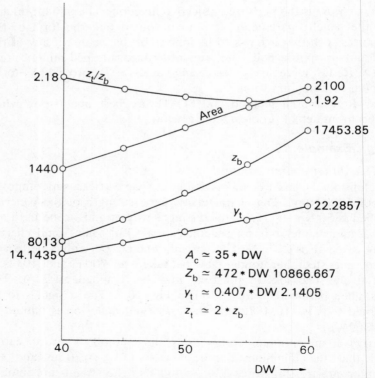

Fig. 10.2 Variation of geometric properties and linearization

Solution
The notation used is given below:

DW—Depth of web in cm
PF—Prestressing force in N
WF—Width of flange in cm
DF—Depth of flange in cm
TWB—Thickness of web at bottom in cm
SL—Slope of web sides
SP—Span of beam in metres
TC—Thickness of topping in cm
FCU—Characteristic cube strength of concrete in N/cm^2
FCI—Cube strength of concrete at transfer in N/cm^2
VC—Allowable shear stress in N/cm^2
VCU—Ultimate shear stress in N/cm^2
ETA—Factor to account for losses in prestress taken 0·78
C_c—Cost of concrete per unit volume in new pence
C_s—Cost of high tensile steel per tonne in new pence
DC—Density of concrete
A_c—Gross area of concrete section cm^2
Z_t, Z_b—Modulus of section with respect to top and bottom fibre in cm^3
FDLT, FDLB—Stresses due to self-weight at top and bottom
FDQT, FDQB—Stresses due to total load at top and bottom
BIU—Ultimate moment due to loads
BIR—Ultimate resisting moment
VCUL—Ultimate shear force
VCO—Ultimate shear force of uncracked section
VCR—Ultimate shear force of cracked section
VCS—Shear taken by auxiliary steel
VP—Vertical component of PF
VU—Ultimate shear stress under loads
VCRM—Minimum allowable shear force of cracked section
DELS—Deflection under service conditions
DS—Density of steel
ECC—Eccentricity of PF in cm
X—Design variable vector
FPU—Characteristic strength of prestressing strands in kN

162 Modern prestressed concrete design

It may be noted that the following are given and are, therefore, preassigned parameters:

WF = 120; DF = 5; TWB = 6·5; SL = 0·10; SP = 12; TC = 5; DC = 2400; DS = 7·84; FCU = 5000; VC = 35; VCU = 530; FPU = 227; $C_c = 700$ p; $C_s = 10\,000$ p. FCI = 3500; ETA = 0·78; $DW_{min} = 20$; $PF_{min} = 40 \times 10^4$

The design variables are just two, DW and PF, and hence direct search is possible.

(1) *Sectional properties*
Sectional properties are worked out for values of DW ranging from 40 to 60 cm. The variation of the properties over this range is graphically presented in Fig. 10.2. It is convenient to linearize the variation of the

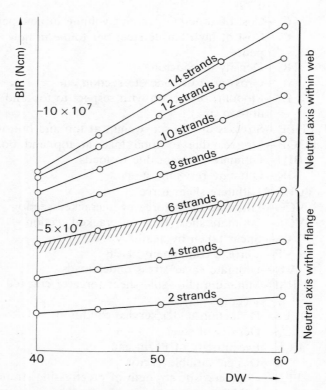

Fig. 10.3 Variation of ultimate resisting moment

properties by the following approximations

$A_c \simeq 35 \times DW$

$Z_b \simeq 472 \times DW - 10\,866 \cdot 667$

$y_t \simeq 0 \cdot 407 \times DW - 2 \cdot 1405$

$Z_t \simeq 2 \times Z_b$

(2) Resisting moment at ultimate

The resisting moment of the section at ultimate is worked out by using the general method outlined in Chapter 4. The variation of the resisting moment for different areas of steel and web depths are presented graphically in Figs. 10.3 and 10.4.

(3) Constraints

(a) Behavioural constraints

At transfer

(i) At supports:

$$\frac{PF}{A_c} + \frac{PF \times ECC}{Z_b} \leq 0 \cdot 50 \, FCI \qquad (10.1)$$

$$-\frac{PF}{A_c} + \frac{PF \times ECC}{Z_t} \leq 100 \qquad (10.2)$$

It may be noted that these constraints at the supports will be inactive, if draped tendons are used as in this example.

(ii) At midspan:

$$\frac{PF}{A_c} + \frac{PF \times ECC}{Z_b} - FDLB \leq 0 \cdot 5 \, FCI \qquad (10.3)$$

$$-\frac{PF}{A_c} + \frac{PF \times ECC}{Z_t} - FDLT \leq 100 \qquad (10.4)$$

Under working loads at midspan:

$$-\frac{ETA \times PF}{A_c} - \frac{ETA \times PF \times ECC}{Z_b} + FDQB \leq 0 \qquad (10.5)$$

$$\frac{ETA \times PF}{A_c} - \frac{ETA \times PF \times ECC}{Z_t} + FDQT \leq 0 \cdot 33 \, FCI \qquad (10.6)$$

$$-DELS + \frac{SP}{360} \leq 0 \qquad (10.7)$$

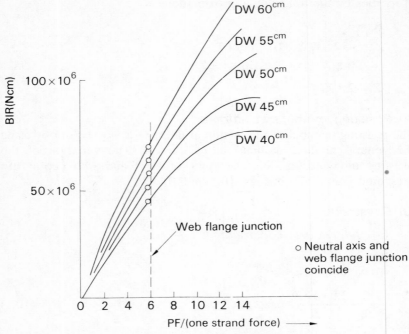

Fig. 10.4 Variation of ultimate resisting moment

At ultimate

$$-\text{BIR} + \text{BIU} \leq 0 \qquad (10.8)$$

$$-(\text{VCO or VCR} + \text{VCS} + \text{VP}^*) + \text{VCUL} \leq 0 \qquad (10.9)$$

$$\text{VU} - \text{VCU} \leq 0 \qquad (10.10)$$

$$\text{VCRM} - \text{VCUL} \leq 0 \qquad (10.11)$$

(b) *Side constraints*

$$-\text{DW} + \text{DW}_{min} \leq 0 \qquad (10.12)$$

$$-\text{PF} + \text{PF}_{min} \leq 0 \qquad (10.13)$$

(4) *Objective function*

The objective function is set up as follows, noting that the cost of prestressing steel and the cost of concrete are the major factors that decide the cost:

$$F(X) = \frac{A_c}{10^4} \times \text{SP} \times C_c + \frac{A_s}{10^6} \times \text{SP} \times \text{DS} \times C_s$$

* VP can be added only to VCO as per Code. For the purpose of computing VP it is assumed that the eccentricity of the draped tendons is zero at the supports.

Optimum design

The cost of stirrups and auxiliary steel in the flange, the cost of formwork and end-formers are the other components of the cost. But these are assumed to be the same for all the candidate designs.

It is also to be noted that in the ultimate limit state the constraint relating to the ultimate bending moment may be critical at or very near the midspan section, and the constraints on shear may be critical anywhere on the span. Hence, these constraints need to be formulated at a number of sections and the most critical included.

(5) Locating the feasible region

The constraints may now be graphically represented (Fig. 10.5). It is clear that the two governing constraints are the limits on stress in the bottom fibre in service and the stress in the top fibre at transfer. The feasible region is shown shaded. Contours of the objective function, which is nearly linear, may now be drawn. The prestressing force PF is not a continuous variable as it can only assume values corresponding to 2 strands, 4 strands, 6 strands, etc. Hence the prestressing forces corresponding to pairs of strands, are to be placed in each rib of the tee, are also drawn in Fig. 10.5. Clearly the point P in the feasible region represents the optimum design and this lies on the cost contour of 1980 p corresponding to 4 strands, two in each rib. If the prestressing force is regarded as a continuous variable, the optimum cost will be on line B at point Q and its value will be 1942 p. The optimum cost of 1980 p corresponds to DW = 50 cm and PF = 545 kN.

(b) The SUMT method

The SUMT method will be explained with the aid of the same example.

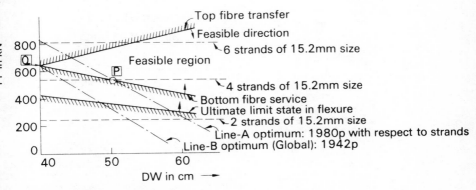

Fig. 10.5 Direct method

Step 1
We start the optimization process at a point in the feasible region with DW = 60 cm and PF = 56 × 10⁴ N. It may be noted that PF has been scaled to be of the same order as DW. That the point lies in the feasible region may be verified by reference to Fig. 10.6.

Step 2
The constraints are evaluated as follows, making use of the known values of DW and PF.

$G(1)$	=	$-783 \cdot 1799$	*Stress in the bottom fibre at transfer at midspan. Inequality (10.3)
$G(2)$	=	$-1 \cdot 3897$	Stress in top fibre at transfer at midspan. Inequality (10.4)
$G(3)$	=	$-124 \cdot 9938$	Stress in bottom fibre in service. Inequality (10.5)
$G(4)$	=	$-1398 \cdot 6920$	*Stress in top fibre in service. Inequality (10.6)
$G(5)$	=	$-444 \cdot 1$	*Deflexion. Inequality (10.7)
$G(5)$	=	$-20\,959 \cdot 020$	*Ultimate moment Inequality (10.8)
$G(7)$	=	$-35 \cdot 5044$	Ultimate shear Inequality (10.9)
$G(8)$	=	$-958 \cdot 8$	*Shear constraint. Inequality (10.10)

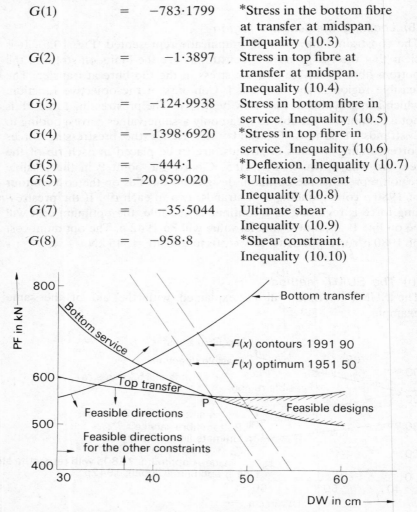

Fig. 10.6 Feasible area for design at midspan

$G(9)$	=	$-137 \cdot 025$	*Shear constraint Inequality (10.11)
$G(10)$	=	-40	*Side constraint on DW Inequality (10.12)
$G(11)$	=	-16	*Side constraint on PF expressed in terms of 10^4 N. Inequality (10.13)
$\sum \dfrac{1}{G_j(X)}$	=	$0 \cdot 7478$	(Excluding inactive constraints which are starred)

Step 3
The initial value of R, designated as R_0, is chosen as

$$\frac{F(X)}{\sum \dfrac{1}{G_j(X)}}$$

so that $\phi(X, R_0) = 2F(X)$
For PF = 560 kN and DW = 60,

$F(X) = 2300 \cdot 50$ p.

Hence, $R_0 = \dfrac{2300 \cdot 50}{0 \cdot 7478} = 3076 \cdot 5$

Step 4
Compute $\phi(X, R_0)$ including all the active constraints.

$\phi(X, R_0) = 4601.$

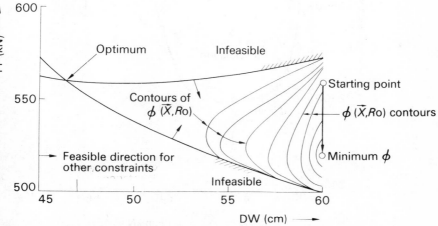

Fig. 10.7 Penalty function approach—first cycle

168 Modern prestressed concrete design

Step 5
The first cycle is illustrated in Fig. 10.7. Compute $\phi(X, R_0)$ for three values of PF, keeping DW = 60. This is known as the method of univariate directions. There are better methods available, such as Powell's Technique and the variable metric method. The method of univariate directions is presented here because of its simplicity. The values of $\phi(X, R_0)$ for the three chosen values of PF are

PF in N	$\phi(X, R_0)$ in p
54×10^4	2579·75
53×10^4	2499·625
51×10^4	2510·60

By quadratic interpolation (Fig. 10.8), the minimum value of $\phi(X, R_0)$ is found as

$\phi(X, R_0)_{min} = 2499·21$ for DW = 60
and PF = $52·2 \times 10^4$ N

Step 6
For cycle 2, $R_1 = 0·80$, $R = 2461$.
Decrease DW to 58.
Repeat the same steps as for cycle 1 to find
$\phi(X, R_1)_{min} = 2370·95$
for DW = 58 and PF = 53×10^4 N

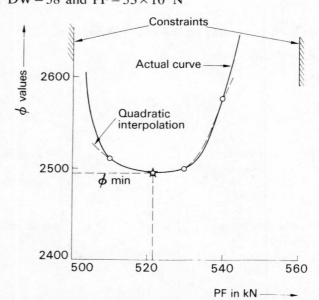

Fig. 10.8 Quadratic interpolation for one move with DW = 60

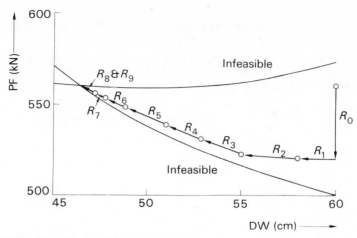

Fig. 10.9 Penalty function approach

Step 7
Continue the process up to cycle 9 as indicated in Fig. 10.9 until the optimum value is reached with $R_8 = 492 \cdot 4$. The optimum value sought is found as $\phi(X, R_8)_{min} = 1951 \cdot 50\,p$

for DW = 46·82 cm and
PF = 55·92 × 10⁴ N.

After the optimum values of PF and DW are obtained, it is necessary to check at sections other than the midspan to make sure that the constraints are not violated. Limits for the cable profiles are shown in Fig. 10.10 for a depth of DW = 50. Similar limits for other depths will be useful in checking final designs.

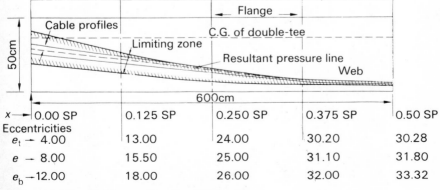

Fig. 10.10 Cable profiles for DW = 50

References

[10.1] Gallagher, R. H., and Zienkiewicz, O. C. (editors), *Optimum Structural Design, Theory and Applications*, Wiley, New York 1973.

[10.2] Ramaswamy, G. S., and Raman, N. V., 'Optimum design of prestressed concrete sections for minimum cost by non-linear programming', paper presented at the Sixth FIP Congress, Prague, June 1970.

[10.3] Fiacco, A. V., and McCormick, G. P., *Nonlinear Programming Sequential Unconstrained Minimization Techniques*, Wiley, New York, 1968.

Index

Absorption, 120
ACI 318:1971, 40, 42, 46, 120
Algorithm, 158
Anchorage, 109, 110, 117
Anchorage End, 5
Araldite, resin, 86
Arches, *See* Prestressing arches

BS 4486, 11
Bars, 4, 5
   deformed, 117, 118
Beam, 73, 97, 119
   Post tensioned, 87, 110
   Prestressed concrete, 36, 40, 43, 47, 137
   Pretensioned, 57
   Rectangular, 36
   Reinforced concrete, 7, 43
   Under reinforced and over reinforced, 36
Beds, *See* Prestressing beds
Bending moments, 97, 125, 128, 133, 135, 141
   Girder, 89
   Slab, 89
Binder, 86
Bridge, *See* Post tensioned bridge construction
British Macallay system, 11
Bridge girder, *See* Post tensioned bridge girder
British code, CP 110:1972, 12, 15, 19, 23, 50, 51

Buckling moments, 30
Bursting zone, 110

Calcium aluminate cement, *See* Cement – calcium aluminate
Cantilever, 142, 143
Cement,
   Calcium aluminate, 2
   Portland, 2
Channel units, 5
Characteristics,
   Loads, 22, 141
   Strength of materials, 24
   Strength of concrete, 24
Chemical, *See* Prestressing, chemical
Classification of concrete structures, *See* concrete structure, classification of
Compression, 96
Compression fibre, 127
Compressive force, 126
   Stress block, 33
Composite section, 89
Coefficient, 16
Computer,
   electronic digital, 157
Concordant, 133
Concordant profile, 134
Concrete structures classification of, 22, 25
Constraints, 157, 158, 163, 165

171

Index

Crack free, 117
  structure, 7
  control, 130
Crack width, 120, 121, 124, 129
Cracking, 27, 43, 51
  web-shear, 41
  flexural shear, 41
Creep, 18–20, 119
  Coefficient, 19
  strain, 20
Curvilinear distribution, 111

Dead load, 23, 25, 148
Deformed bars, See Bars, deformed
Degree of prestress, 20
  Index, 131
  Expanding cement, 2
  Expanding concrete, 2
  Hydrothermal method of curing, 2
Deflextion, 130
Depth factor, 121
Design
  flexure, 27, 30, 38, 43
  shear and torsion, 40
  shear reinforcement, 44
  post tensioned member, 86
  pretensioned products, 54
  end block, 107
  limit states, 22
Design load, 23, 117
Design variables, 158
Direct search method, 158, 159
Ductile, 120

Eccentricity, 66, 67, 93, 100, 136, 140
Elastic compression, 14, 15
Elastic material, 6
Elasticity, See Modulus of elasticity
Electro-thermal method, 3, 120

Electro-thermal prestressing plant, 5
Electronic digital computer, See computer, electronic digital
End blocks, 109, 110, See also Design, end blocks
Epoxy mortar, 86
Equivalent loads, 134, 135, 137
  concept of, 138

Fibre, 90, 94, 100
Fibre stress, transfer
  working loads
  top, 29
  bottom, 29
Filler, 86
Flexural cracking zone, 41
  See also Design, flexure
Flexural crack, See design, flexure
Flexural, shear cracking, See cracking, flexural shear
Flexural tensile, 116
Fexure, 97
  See also Design, flexure
Friction, 16
Force, See Prestressing, force

German Dywidag system, 11
Girder, 96, 97
Global safety factor, See Safety factory
Grouting, 86
Gypsum (plaster of Paris), 2

Heterogeneous materials, 6
High strength deformed bars, 4
High tensile alloy rods, 12
High tensile wires, 2
High yield strength, 118
Humidity, 18
Hydraulic jacks, 5
Hydraulic prestressing equipment, 4

Hyperbolic curve, 57, 58
Hyperbolic paraboloid See Shell, hyperbolic paraboloid
Hyperboloid, 58, 59
Hypothetical tensile stress, 120, 121

Indeterminate structures, 133
Interface, 106, 107

Limit State Design, See Ultimate Limit State
Linear Programming Problem, 158
Linearization, 158, 159
Load, See Characteristic, loads
Load deflexion curve, 120, 131
Low-heat treatment, 9, 11

Mechanical, See Prestressed Mechanical
Midspan, 67. 69. 82, 135, 137
Modulus of elasticity, 18, 20
Modulus of rupture, 25
Moment of intertia, girder, 88
  slab, 89
Magnel inequalities, 28, 65
  equation, 62
  chart, 30
  diagram, 67
  camber, 79
Moderate Prestressing, 119
Merit function, 157

Non-linear programming problem 158

Optimization, 157, 158, 166
Optimum design, 157, 159

Parabola, 57, 142
Parabolic cable, 135
Parabolic profiles, 98, 135, 136

Parameters, 157
Partial prestressing, 119
Patented coil, 9
Plaster of Paris, See Gypsum
Portland cement, See Cement, Portland
Post tensioned beam, See Beam, post tensioned
Post tensioned bridge construction, 86
Post tensioned bridge girder, 87
Post tensioned members, See Design, post tensioned members
Post tensioning, 2, 14, 19, 21, 86
Precompression, 1, 2, 7
Pressure pipes, 2
Prestressed poles, 3
Proof stress, 11, 12
Penalty function
  interior, 159
  exterior, 167
  approach, 169
Preassigned parameters, 157, 159
Precast trusses, 3
Pressure line, 30, 133, 140
Prestress, 1, 97, 116, 117
  losses in, 14, 21, 81
Prestressed, 116
  beam, 7, 120, 133
  concrete, 1, 13, 51, 116, 117
  definition, 1
  structure, 7, 25, 117, 133
  mechanical, 5
  roofing elements, 5
  structures, 7, 25, 30, 119
  poles, 3
Prestressed concrete beams, See Beam. prestressed concrete
Prestressed concrete bridge, 116
Prestressed design, 117
Prestressing, 1, 41, 116, 138
  advantage of, 7
  arches, 1

Prestressing (*cont.*)
  beds, 9
  chemical, 2, 3
  force, 29, 47, 66, 82, 90, 100, 126, 135, 142, 159
  steel, 48, 117, 118, 119
  strand, 10
  tendons, 11, 42, 46, 55, 97, 127, 130, 134, 142
  wires, 2, 9, 10
Prestressing losses, *See* Prestress, losses in
Prestressing steel, *See* steel, prestressing
Prestressing tendons, *See* prestressing, tendons
Prestressing wire, 3
Pretension, 5
Pretensioned beams, *See* Beam, pretensioned
Pretensioned girder, 124
Pretensioned hyperboloids, 54, 57
Pretensioned products, *See* Design. pretensioned products
Pretensioning, 2, 20
Principal tensile stresses, 116

Quartz sand, 86

Railway sleeper, 3, 14
Rectangular beams, *See* Beam, rectangular
Reinforced concrete, 1, 116, 117, 119
Reinforced concrete beams, *See* Beam, reinforced concrete
Reinforced concrete members, 117
Reinforced concrete structure, 7, 25
Resisting moment, 163
Roofing elements, *See* Prestressed, roof elements
Rupture, *See* Modulus of rupture

Safety factor, 23, 24
Segment, 86
Shear, 72, 97
Shear and torsion, *See* Design, Shear and torsion
Shear force, 99, 135
Shear reinforcement, 44, 47
  See also Design, Shear reinforcement
Shear resistance, 50
Shear stress, 47, 51, 84, 85, 97
Sheathing, 117
Shell, 55, 59, 155
  Hyperbolic paraboloid, 3
  silberkuhl, 54
Shrinkage, 1, 18, 20, 27
Silberkuhl shell, *See* Shell, silberkuhl
Silica, 86
Slab, 96, 97, 154
Steam-curing cycle, 13
Steel, 95
  Prestressing, 48, 127, 128, 159
  Tensioned and untensioned, 33, 34, 46, 117, 120, 121, 124, 127
Stirrups, 99, 102
Strain, 95, 116
Strength of concrete, *See* characteristic, strength of concrete
Strands, 92, 126
  See also prestressing, strand
Strength of materials, *See* Characteristic strength of materials
Stress, Strain curve, 13, 34, 35, 127
Stresses, 3, 67, 70, 82, 83, 90, 97, 116
Structural analysis, 157
Structures, *See* Prestressed, structures; Reinforced concrete, structures

Surface tension, 18
Swiss Code SIA 162:1968, 46, 50
Symmetrical prisms, 109
SUMT, 158. 159, 165

Tendons, 2–4, 10, 14, 15, 30, 81, 86, 90, 92, 98, 100
  See also Prestressing tendons
Tensile brusting force, 111
Tensile deformed bars, 118, 120
Tensile force, 44, 111
Tensile strength, 27, 40, 42, 112, 123
Tensile stress, 1, 9, 25, 27, 41, 107, 118, 119, 121
Tension, 37, 84, 96, 112, 116
Tension fibre, 120
Tension force, 120
Tensioned and untensioned steel, *See* steel, tensioned and untensioned
Tensioning, 2
Torsion, 51
Transverse stress, 108
Transition, 116
Transverse bending moment, 156
Transmission length, 3, 78
Trusses *See* precast trusses

Ultimate bending moment, 95, 101
Ultimate flexural resistance, 97
Ultimate limit state, 22, 24, 27, 32, 40, 47, 97, 118, 126
Ultimate loads, 42
Ultimate moment, 95, 96, 128
Ultimate shear, 46, 73, 84, 101
Ultimate strength, 12, 22, 36, 72, 84
Under reinforced and over reinforced Beams, *See* Beam, under reinforced and over reinforced
Unified code, 22
Untensioned reinforcement, 118, 120
Untensioned steel, *See* Steel, tensioned and untensioned

Water cement ratios, 14, 18
Web, depth of, 159
Web depths, 163
Web, shear cracking, *See* Cracking, web shear
Wires, *See* Prestressing wires
Wobble, friction, 16
  coefficient, 16

Yield strengths, 3